THE NATURE OF
NORTHERN AUSTRALIA

Natural values, ecological processes and future prospects

1. *(Inside cover)* Lotus Flowers, Blue Lagoon, Lakefield National Park, Cape York Peninsula. *Photo by Kerry Trapnell*

2. Northern Quoll. *Photo by Lochman Transparencies*

3. Sammy Walker, elder of Tirralintji, Kimberley. *Photo by Sarah Legge*

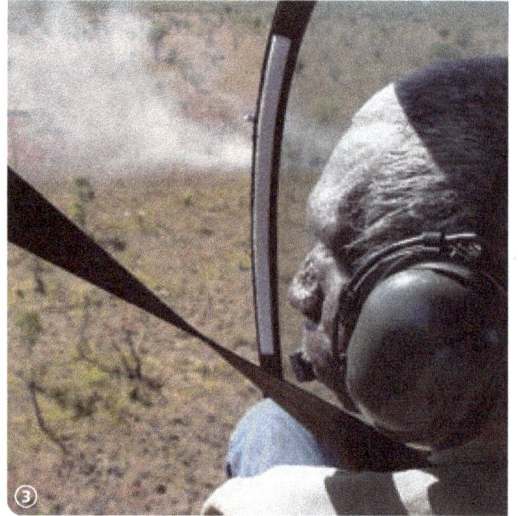

4. Recreational fisherman with barramundi, Gulf Country. *Photo by Larissa Cordner*

5. Tourists in Zebidee Springs, Kimberley. *Photo by Barry Traill*

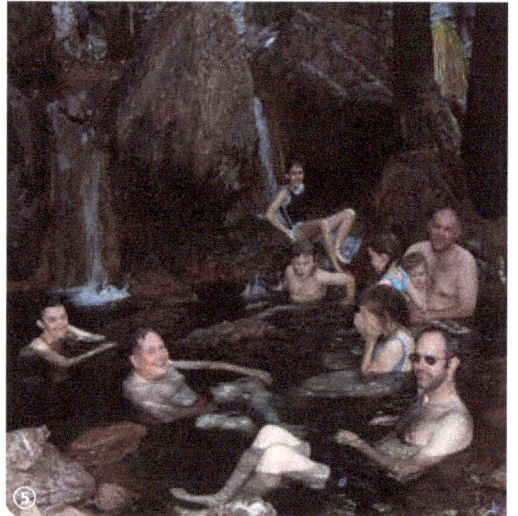

6. Dr Tommy George, Laura, Cape York Peninsula. *Photo by Kerry Trapnell*

7. Cattle mustering, Mornington Station, Kimberley. *Photo by Alex Dudley*

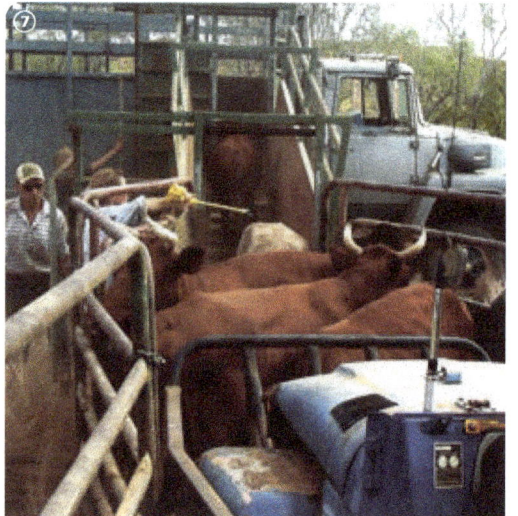

THE NATURE OF
NORTHERN AUSTRALIA
Natural values, ecological processes and future prospects

AUTHORS John Woinarski, Brendan Mackey, Henry Nix & Barry Traill

PROJECT COORDINATED BY Larelle McMillan & Barry Traill

THE AUSTRALIAN NATIONAL UNIVERSITY

E PRESS

TROPICAL SAVANNAS CRC
Cooperative Research Centre for Tropical Savannas Management

THE WILDERNESS SOCIETY

THE MYER
FOUNDATION

Charles Darwin
UNIVERSITY

ANU

E PRESS

Published by ANU E Press
The Australian National University
Canberra ACT 0200, Australia
Email: anuepress@anu.edu.au
Web: http://epress.anu.edu.au
Online version available at: http://epress.
anu.edu.au/nature_na_citation.html

National Library of Australia
Cataloguing-in-Publication entry

The nature of Northern Australia:
it's natural values, ecological
processes and future prospects.

ISBN 9781921313301 (pbk.)
ISBN 9781921313318 (online)

1. Environmental sciences - Australia,
Northern. 2. Human ecology - Australia,
Northern. 3. Social ecology - Australia,
Northern. 4. Human geography - Australia,
Northern. 5. Australia, Northern - Economic
conditions. 6. Australia, Northern -
Social policy. I. Woinarski, John.

333.70994

Printed by University Printing Services, ANU

This edition © 2007 ANU E Press

Cover photo: Eucalypt Savanna in Kakadu
National Park by Glenn Walker.

Design by Oblong + Sons Pty Ltd
07 3254 2586
www.oblong.net.au

THE DOTS

We've used as the basis for the book two
of the central findings of the book itself,
namely that the ecosystems across the whole
of the North are inextricably intertwined
and that their conditions vary greatly from
season to season and from year to year.

To convey this we've chosen six representative
centres – Broome, Kununurra, Darwin,
Katherine, Normanton and Weipa – which cover
all of Northern Australia west to east and which
vary north to south and between the coast and
inland. The dots at the bottom of each page
represent these six centres. The dots change
colour and number according to the land and
weather conditions at each centre across the
course of a chosen year, giving a subtle indication
of the inter-related weather patterns across the
whole of the North and the variability of the
conditions. The colours of headings, captions
and pull-out boxes change per chapter as the
weather patterns cycle through, corresponding
with the dots at the bottom of the page.

CONTENTS

1 Waterbirds, Aurukun Wetlands, West Coast Cape York Peninsula. *Photo by Kerry Trapnell*

ACKNOWLEDGEMENTS

The authors and contributors to this report acknowledge the Indigenous Traditional Owners of Country throughout Northern Australia, and their right to speak for Country.

The authors are recipients of an ARC Linkage Grant focused on investigating connectivity conservation issues in various regions, including Northern Australia. Some of the outcomes from that research were drawn upon for this report.

The authors would like to thank the following people and organisations for their assistance with the publication:

The Myer Foundation for its generous funding, without which the project would not have been possible.

The Dara Foundation for its support of the establishment and maintenance of the WildCountry Science Council.

Larelle McMillan and Lena Ruby for their support in bringing this project together.

Written contributions from Alan Andersen, Sandra Berry, Peter Jacklyn, Stephen Garnett, Raina Plowright, Samantha Setterfield and Dick Williams.

Our reviewers – Peter Cooke, Stephen Garnett, Bob Wasson, Alex Kutt, Garry Cook, Chris Margules, Peter Jacklyn, Rosemary Hill, Stuart Blanch, Sarah Legge, Paddy O'Leary and Peter Whitehead.

All photographers who have generously contributed to the publication (acknowledged individually in the text).

Andrew Wong and Luciana Porfirio for producing the figures and graphs.

Anthony Esposito, Peter Jacklyn, Joe Morrison, Michael Soulé, Jann Williams, Cecile van der Burgh, Andrew Wong, Simon Kennedy and James Watson for additional comments on what the project should cover and specific text.

Julie McGuiness, Christabel Mitchell, Lyndon Schneiders, James Watson, Virginia Young, Heidi Douglas and Sylvia Tobler for support in different ways throughout the project.

The Tropical Savannas Cooperative Research Centre for assistance, and for supporting some of the research on which this book is based.

Lorena Kanellopoulos and ANU E Press for their assistance in publishing this report.

Our editors, Hilary Cadman and An Van den Borre, for their bringing together of the authors' four writing styles.

And lastly thanks to everyone at Oblong + Sons for their skill and patience with design and printing of the publication.

PREFACE

WHY DID WE WRITE THIS BOOK?

It is because we see something of value here, that is in the process of being devalued. It is because we are aware of the increasing momentum to transform these lands, largely by those unaware of or impervious to its intrinsic value. It is because we consider that the fate, health and societies of people in Northern Australia are bound inextricably with their natural landscapes. It is because it is time to consider systematically and realistically the future of this land, and mechanisms to guide it to that future. It is because of the failures of the past.

We take inspiration and instruction from our history. Notwithstanding a general anarchic pattern of development, there have been some extraordinarily detailed and systematic attempts to understand the North and to plan its future. The most grand of these was the North Australia Development Committee which, from 1946 to 1948, was charged with investigation of the lands and their potential for pastoral, agricultural, mining, forestry, marine, fuel and power, and processing and manufacturing industries. Thirty years later, its chairman, Nugget Coombs, reflected on its limitations and biases:

Apart from its emphasis on research and experiment, the committee's work reflected the optimism of the time and the prevailing views, shared firmly by the committee itself, that growth was a good thing, that it could be achieved primarily by seeking to impose on the North a pattern of productive activity and a way of life essentially European in its origin and substantially European in its relevance. There is little evidence in the work of the committee, I am sorry to say, of a genuine understanding of, or an intuitive sympathy with, the climatic and territorial environments of the North. We were southerners, and Europeans, and never really got over the sense of being in a somewhat alien and hostile environment ... There was an almost complete disregard for the character and potential of the natural flora and fauna of the North ... There was nothing organic in the growth we planned for; it was fundamentally to be based on extractive and exploitative techniques ... It is this lack, not merely in the work of the committee, but in the thinking about developments by me and my colleagues and the generation to which we belonged, which in retrospect is so disappointing. (Coombs 1977)

Thirty years further down the track, it is timely that we mark this message, accept the shortcomings of such narrowly developmental perspectives, and work towards a future for Northern Australia that recognises and nurtures its societies and land, and the links between them. It is in this spirit that this book is offered.

INTRODUCTION

If you leave the Wet Tropics around Cairns, and head west by car for an hour or so, the road goes up and over the mountains that lie behind the coast. On the other side, the rainfall drops off quickly, and you enter the great 'sea' of savanna that stretches across Northern Australia. Still heading west, several long days of driving later, you will reach Broome on the edge of the Indian Ocean.

For all this time, for all the 3000 or more kilometres of travel, you will have been among vast areas of eucalypt savannas and native grasslands, broken only by an occasional cleared paddock, a scattering of small towns, and the rivers and wetlands that give life to the country.

This landscape of savanna and rainforest, rivers and wetlands, is of great significance. On a global scale, such large natural areas are now very rare. Northern Australia stands out as one of the few very large natural areas remaining on Earth: alongside such global treasures as the Amazon rainforests, the boreal conifer forests of Alaska, and the polar wilderness of Antarctica.

Unlike much of southern and eastern Australia, nature remains in abundance in the North.

Great flocks of birds still move over the land searching for nectar, seeds and fruit. Rivers still flow naturally. Floods come and go. In fertile billabongs, thousands of Magpie Geese, brolgas, egrets and other water birds still congregate.

The intact nature of the North provides a basis for much of the economic activity and the general quality of life for residents of the area. Most of the major industries – tourism, pastoralism, Indigenous economies – rely on productive, functioning and healthy natural ecosystems. Across the North, recreational activities such as fishing, four-wheel driving and visiting beautiful country depend on the opportunities provided by a largely intact and natural landscape. Being in and among nature remains a normal part of life for people in the North, in contrast to the situation for those living in the now highly transformed, cleared and urbanised areas of southern Australia.

For the high proportion of Northern Australian residents who are Indigenous, country is part of the essence of life. Knowledge of and links to the land remain strong, and there remains an enduring responsibility to look after the land, and its plants and animals.

However, there are increasing signs that much of the country in Northern Australia is not healthy. Some native plants and animals are declining, and many non-native plants and animals are increasing. Many people in the North live in communities with poor physical, social and economic health. There is increasing debate and conflict over the best use of natural resources.

WHAT IS A BIOREGION?

A bioregion is an area defined by a mixture of biological and geographical factors (Thackway and Cresswell 1995). For example the Central Kimberley Ranges are often defined as a distinct bioregion. The region has characteristic rocky sandstone ranges that have scattered trees and spinifex grass. The ranges are separated by valleys that support eucalypt savanna on deeper alluvial soils. The climate is monsoonal, with low annual rainfall tending towards semi-arid.

OBJECTIVES OF THIS STUDY

This study arose because of the authors' concern about the future of the North's natural values. A comprehensive account of these values is long overdue. Over the last decade there have been great advances in understanding the ecological workings of the North. But for most Australians, the North remains a strange and somewhat foreign land, stereotyped by myth and tourist advertisements. Planning for the future of the North, or indeed of Australia as a nation, cannot rest on such a shallow base.

We have written this report for all those concerned with the long-term future of the North.

We share the view that given the significance of its natural values, the desirable future we should consider for Northern Australia is one that looks after the country, its people and its wildlife.

FIGURE 1.1 MAJOR REGIONS OF NORTHERN AUSTRALIA

In this study, we report on three specific sets of knowledge and analysis:

- An evaluation of the natural values of Northern Australia, drawing on contemporary environmental, ecological and conservation science;
- An assessment of 'how the country works'; that is, the ecological processes that maintain country, wildlife and people in Northern Australia;
- Building on the first two points, an analysis, set of recommendations and advice on how best to protect country and wildlife, and to lay the foundation for healthy communities and people in Northern Australia for the long term.

THE REGION

Northern Australia is not a political entity (although there have been occasional attempts to establish such a state), nor is it a rigidly demarked environmental unit. Across much of the North there is a gradual transition from high-rainfall coastal areas (with denser vegetation) to semi-arid and arid inland Australia (with more open vegetation). Hence, any bounds to the region are partly artificial. Historically, there have been a range of attempts to delineate 'Northern Australia'. One commonly used boundary is that of the Tropical Savannas Cooperative Research Centre (CRC); this was largely based on a conventionally accepted national mapping of bioregions.

This study broadly follows the CRC definition and boundaries, except that we exclude from most considerations the Central Mackay Coast, part of the Mitchell Grass Downs, Wet Tropics and the Brigalow Belt North bioregions, along the southern and eastern flanks of the area defined by the CRC. We also deliberately exclude the Wet Tropics bioregion – Australia's tropical rainforest core, around Cairns – because its climate, topography and environments are notably different, and because it has received relatively intensive study elsewhere.

Thus, our study region comprises those areas generally known as the Kimberley, Top End of the Northern Territory, the Gulf Country and Cape York Peninsula, extending south into north-central Queensland, around and including the Einasleigh Uplands (Figure 1.1). This covers an area of more than 1.5 million square kilometres. As so defined, Northern Australia includes all of the catchments of rivers flowing into the northern tropical seas – the Indian Ocean, Joseph Bonaparte Gulf, Timor Sea, Arafura Sea and Coral Sea, and includes lands most exposed to a monsoonal climate.

Not included in this study are the marine ecosystems of the North. These range from the Great Barrier Reef in Queensland to the seas around the Buccaneer Archipelago on the Kimberley Coast. Ideally, this study would have included as much information on marine ecosystems as on terrestrial systems. However, the information available on most marine systems in Northern Australia remains scant and patchy. There are still major gaps in knowledge on the ecological connections and processes that operate in northern seas, and of the processes and patterns that link the marine and terrestrial environments. We therefore leave marine issues for a future study. Our study of Northern Australia extends to the estuaries and mangroves, but no further.

INDIGENOUS KNOWLEDGE AND VALUES

This study reflects the perspective of western-trained conservation scientists. It does not present an Indigenous perspective, nor Indigenous ecological knowledge of the North. As such, we acknowledge that this story remains incomplete.

Indigenous laws and customary practices have shaped the environments of Northern Australia for thousands of generations, and change from this traditional management has contributed to the ecological problems now facing many parts of the North.

In considering the future of Northern Australia's biodiversity and associated natural values, Indigenous ecological knowledge, both traditional and contemporary, and Indigenous land management have an essential and distinctive role. A large proportion of Northern Australia is owned and managed by Indigenous people, and there remain strong attachments of Indigenous people across the entire landscape.

Much of the fate of the North's environments is inexorably linked to the cultural integrity and prosperity of Indigenous societies. Many Indigenous communities are beset by poor standards of health and education, by poverty and by cultural decay; and this is reflected in and further fuelled by deterioration of Indigenous-owned lands and their natural values.

Environmental management is one area that offers hope, employment and cultural respect for Indigenous communities. As evident in some examples in this book, the North is at the forefront of such management, and there are an increasing number of successful cases – of joint management of conservation reserves, of Indigenous Protected Areas, of Indigenous ranger groups, of use of traditional fire management for carbon trading – each demonstrating some connection between healthy lands and healthy communities. While such examples fit readily into the theme of this book, we recognise that we do not speak for Indigenous aspirations, and that at least some Indigenous peoples will have very different perspectives on the natural values in Northern Australia.

THE STRUCTURE OF THIS STUDY

In the study we begin with describing the North and how its environment works, and finish with a framework for the future. The specific chapters are:
- A general description of features of Northern Australia, the country and its people (Chapter 2);
- An account of the ecological processes and connections that maintain the area's natural values (Chapter 3);
- A summary of specific natural values of the North, where possible within a national and international context (Chapter 4);
- An assessment of factors that may threaten the natural values and ecological processes of the North (Chapter 5); and
- A framework that provides for the ongoing maintenance of the values of Northern Australia (Chapter 6).

Where relevant, we have supplemented the text with boxes that expand on particular themes.

THE AUTHORS

The authors are members of the WildCountry Science Council, an independent volunteer group of leading scientists who provide advice on the large-scale, long-term requirements for biodiversity conservation in Australia. The Science Council identified Northern Australia as a priority issue and decided it would be timely to bring together current knowledge about the region's natural values and the underlying ecological processes.

The chief authors of this study are all ecological and environmental scientists who have worked on and in Northern Australia for many years. Our perspectives and expertise reflect our backgrounds in research into ecological issues and our work on projects that have sought conservation outcomes that work for people and country. We all support priority being given to protecting and maintaining natural ecosystems as the basis for a sustainable Australia. The organisations and institutions we work for are diverse and include the Australian National University, Northern Territory Department of Natural Resources, Environment and the Arts, the Tropical Savannas CRC and The Wilderness Society. The views expressed here are entirely those of the authors and are not necessarily the views of these organisations.

THE LAND AND ITS PEOPLE

The environment of Northern Australia can seem an odd mix. While it has an intimate familiarity to local Indigenous people, to those accustomed to temperate Australia, it has a strange character. Fires seem too pervasive and frequent; many of the native trees are at least semi-deciduous (they lose their leaves to save water during the Dry season); there is too much grass, some of it taller than a person; the eucalypts don't have that familiar evocative reassuring smell; even the colours of the bush seem somewhat harder. Parts of the landscape seem decidedly African in flavour, with the boab trees but without the lions. Indeed, the link across the Indian Ocean is explicitly marked: the Kimberley region is named for its resemblance to the landscape in southern Africa of the same name.

Why does the North appear so different, such a mix of familiar Australian and apparently non-Australian natural features? In large part, it is an artefact of settlement history, of where most Australians grew up. The distinctive features and feel of Australia are dominated by the environments that characterise where most people live – coastal southern, eastern and south-western Australia and their agricultural hinterland.

Earlier settlers in southern Australia may have found the native vegetation there strange too but were able to transform it to a more homely fashion, with European plants and animals. That has not been the case in the North; for most Australians, it remains a foreign and unfamiliar landscape, and even its society seems different.

In this chapter we provide an overview of the North's geography and consider how it differs from the rest of Australia. We also introduce its regions, weather, landscape features and people. Finally, we discuss the North's current land tenure and economies, as well as the way in which the land is managed. These facts and figures provide the background necessary to understand the natural values, processes and threats, discussed in the following chapters.

THE PAST: AN ARAFURA INHERITANCE

In part, Northern Australia's legacy now lies submerged. For much of the past three million years, Australia and New Guinea were a single land mass, with a wide plain across what is now the Arafura Sea. The only high ground on the plain were low hills that are now islands

1 Boys fishing, Elin Beach, Hopevale, Cape York Peninsula. *Photo by Kerry Trapnell*

① Dr Tommy George, Laura, Cape York Peninsula. *Photo by Kerry Trapnell*

to riparian zones and protected gorges in the south, with much of the Arafura plain a savanna, similar to parts of Northern Australia today (Nix and Kalma 1972; Schulmeister 1992).

The eucalypt savannas that now dominate the North probably began developing from 15 million years ago as the continent's climate dried. From three million years ago the dominance of savanna accelerated as the periods of massive global cooling created drier conditions (van der Kaars 1991). This directional change may have been further influenced by the arrival of Aboriginal people and consequential altered fire regimes, some 50,000 years ago (Veevers 2000).

The alternating global warming and cooling episodes of the last three million years repeatedly isolated then reconnected New Guinea and Australia, as the Arafura plain successively submerged and emerged with changing sea levels, as show in Figure 2.1 (Voris 2000). The last glacial maximum was approximately 20,000 years ago. As Earth warmed and ice melted, the land bridge was submerged (approximately 10,500 years ago) and the final connection across Torres Strait eventually flooded (8000 years ago), and Australia was isolated. The modern shoreline of Australia dates from about 6000 years ago (Chapell and Thom 1986; Lees 1992).

One of the features of the Arafura plain before the last sea-level flood was Lake Carpentaria – a giant (~30,000 km²: Torgersen *et al.* 1988) freshwater lake in the centre of what is now the Gulf of Carpentaria. As the sea waters rose, this lake disappeared. One reminder of this ancient lake is the current fragmented distribution of fish such as the Threadfin and McCulloch's Rainbowfishes, Lorentz's Grunter and the giant Freshwater Anchovy in rivers of Arnhem Land, Cape York and southern New Guinea. These rivers all previously intermingled in Lake Carpentaria.

The rising sea also fragmented the range of many other plants and animals. Comparable environments and species assemblages persist in the Fly River region, Port Moresby and Popondetta areas of southern New Guinea, and across Northern Australia. The connections were especially strong, close and more enduring between Cape York Peninsula and New Guinea. Cape York still harbours most of the species shared between Australia and New Guinea, especially rainforest species (Nix and Kalma 1972).

fringing the Kimberley coast and Arnhem Land, the islands in Torres Strait and the low hills that fronted the north-western coastline of the Arafura plain (now the Aru Islands of Indonesia).

Major river systems flowed across this plain, arising from both the south and the north. The plain had vast shallow lakes, and embayments fringed with mangroves and salt-marsh. Rainforests were largely confined to the mountains and slopes to the North and

The consequences of these dramatic shifts in sea level and associated climatic and vegetation changes must have been profound for Aboriginal people. Thousands of generations watched sea levels rise and decline. For much of that time, clan territories could have extended across the Arafura plain. The marshy shores of Lake Carpentaria, now well out to sea, may once have been productive and well-populated lands. The memory of these lost lands is still evident in some coastal Aboriginal cultures.

THE NORTHERN LANDSCAPE

Across Northern Australia there are striking similarities over vast areas of country, in climate, landforms, vegetation and wildlife. Often the similarities dominate the casual observer's perspective. On Cape York Peninsula you can stand on deep red laterite soil and admire the surrounding tall forests of the Darwin Stringybark *Eucalyptus tetrodonta*. Thousands of kilometres away, in Arnhem Land or the north Kimberley, you can be in almost identical stands. The North has a distinctive feel, and is an environmental entity arising from the characteristic features of climate, landforms, vegetation, wildlife and ecological functioning.

CLIMATE

Much of the present character and ecology of the North is defined by a strongly monsoonal climate, a feature it shares with tropical environments in Asia, Africa and South America. The land is exposed to an orderly procession of climatic extremes: an almost rainless Dry season of about 7–8 months, followed by a shorter season of violent storms and torrential rains. Temperatures are high year-round, peaking in the summer Wet season. Cyclones are frequent. It is a harsh climate that shapes the ecology, distribution and abundance of most plant and animal species, and the interactions between them. The climate is also a major constraint on the kinds of land use activities that the North can sustain, and has been a major force shaping Indigenous land management and cultural practice.

There are distinct geographic patterns in climate (Figure 2.2). Annual rainfall decreases away from the coast towards Australia's arid core. Rainfall is highest on the east coast of Cape York Peninsula and the northernmost coastal areas of the Northern Territory and the Kimberley.

FIGURE 2.1 **SHIFTING PATTERN OF NORTHERN AUSTRALIA/NEW GUINEA COASTLINES**

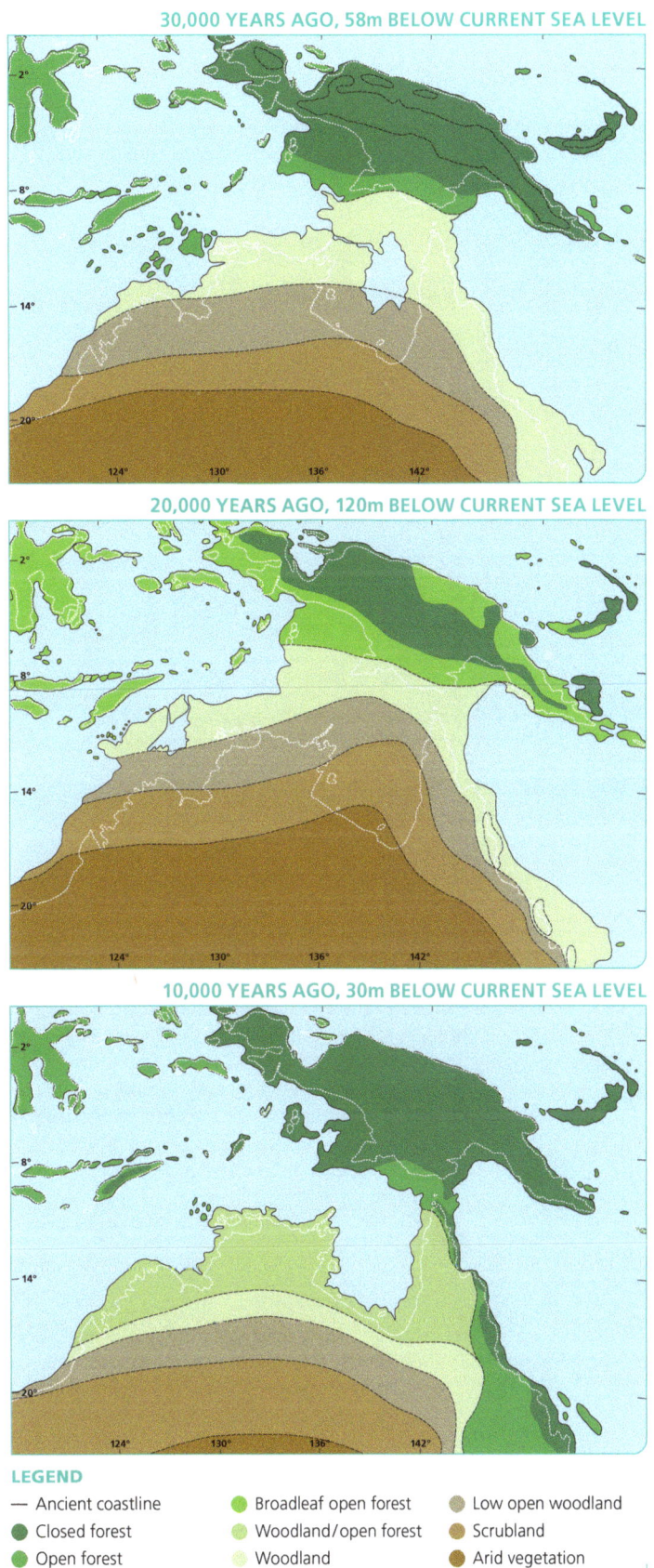

30,000 YEARS AGO, 58m BELOW CURRENT SEA LEVEL

20,000 YEARS AGO, 120m BELOW CURRENT SEA LEVEL

10,000 YEARS AGO, 30m BELOW CURRENT SEA LEVEL

LEGEND

— Ancient coastline
● Closed forest
● Open forest
● Broadleaf open forest
○ Woodland/open forest
○ Woodland
● Low open woodland
● Scrubland
● Arid vegetation

Source: Nix and Kalma 1972.

9

SOILS

Much of Northern Australia lies on ancient rocks which arose from different geological episodes over the past three billion years. First, the southern half of Western Australia emerged and this formed the basis of the Australian continental plate. Second, about two billion years ago, an accretion of basement and platform cover formed the western two-thirds of Northern Australia (the western margins of the Gulf Region, Top End and Kimberley). Third, the Cape York basement and uplands emerged at around 500–200 million years ago.

Subsequently, the extensive intervening sediments of the Great Artesian Basin and central lowlands began developing about 300 million years, reached a maximum 135–65 million years ago and continue today. Thus, Northern Australia is home to ancient rocks and ancient landscapes that have been subject to millennia of deep weathering.

From this parent material, thin and infertile soils have developed, leached of nutrients by intense tropical rains over millennia. Only in a few districts are there younger and more fertile soils, developed from volcanic rocks, limestones or floodplain alluvium. The cracking clays on these soils support extensive grasslands, and low, open grassy woodlands. Vast areas across the whole of the North are mantled by red and yellow earths and deep sands that support grassy woodlands or savannas. Equally vast are areas of shallow or skeletal soils, often sandy, that support hummock grassland (spinifex) and heathland.

High rainfall intensities, particularly during the first storms of the Wet season, have great erosive power. Consequently, even in undeveloped, ungrassed landscapes there is continuous loss of soil by sheet wash and very little profile development on slopes greater than three per cent.

LANDFORMS

Many visitors to the North see the land as largely featureless – vast expanses of bush on flat lands extending to the horizon. Indeed, much of the North is gently undulating land at low elevation. From the southern Kimberley, east to Cape York Peninsula these landscapes are unbroken by high mountains or ranges, and landscape variation is subtle. A consequence of the region's low topographic relief is that savanna species have been able to disperse across much of the North, unhindered by mountain barriers, for millions of years.

However, there are substantial areas of far more complex landforms. These areas provide much of the dramatic attraction to visitors, and harbour many of Northern Australia's most distinctive endemic plant and animal species. The spectacular ranges and escarpments of the Kimberley, the Ord-Victoria uplands in the southwest of the Top End, Arnhem Land and the uplands of the western Gulf

country in Queensland are examples. The ruggedness of these features is notable on a national scale (Figure 2.3): the Kimberley has a more rugged topography, with more cliff lines, than any other part of Australia.

This concentration of cliffs, escarpments and rocky areas has favoured a proliferation of distinctive rock-dwelling mammals (Freeland *et al.* 1988) and other animals and plants (e.g. Woinarski *et al.* 2006a). It has also provided the necessary environment for the development of the richest rock art galleries in the world.

However, these spectacular landscapes are of modest elevations (few peaks exceed 900 m) and while they do have orographic effects on regional rainfall they do not modify the underlying seasonally wet-dry climate. Only mountains in the far north-east are high enough to intercept the easterly trade winds and generate orographic rainfall in the winter Dry season (e.g. the McIlwraith Range on the east coast of Cape York Peninsula).

FIGURE 2.2 **AVERAGE ANNUAL RAINFALL**

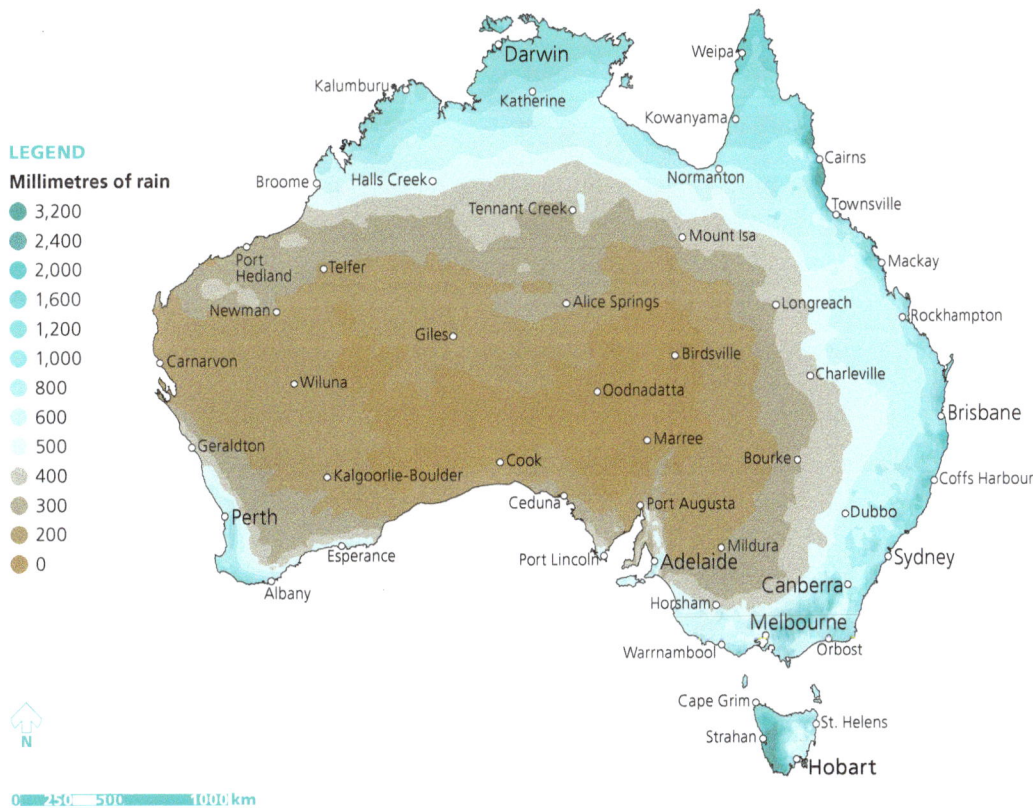

LEGEND

Millimetres of rain

- 3,200
- 2,400
- 2,000
- 1,600
- 1,200
- 1,000
- 800
- 600
- 500
- 400
- 300
- 200
- 0

Source: Based on a standard 30-year climatology (1961–1990), Bureau of Meteorology.

ENVIRONMENTS

One feature of the vegetation across the North is the widespread distributions of many savanna plants. These species evolved and spread because they had few climatic or other isolating barriers. In contrast, rainforest and some heathland species have faced long periods of isolation, persisting in patches of particular soil, geology or landform, physically isolated in a sea of savanna. Consequently, many locally endemic species have evolved in habitats such as rainforests and on isolated sandstone escarpments.

The following sections describe some of the key species and vegetation types of the North.

Savanna

One of the results of the monsoonal climate and relatively infertile soils is the dominance of a single broad vegetation type – savanna – from the Kimberley to Cape York, covering more than 1.5 million square kilometres (Figure 2.4).

Savanna is the result of highly seasonal rains and the regular fires of the long Dry season. The grasses include annuals and perennials, and both types have short, intense growing periods during the Wet season, then either die or dry off to the roots during the Dry. Many grasses respond positively to fire, resprouting rapidly after burning.

Eucalypts are the dominant trees in the Australian savannas. These are tolerant of the annual drought and usually highly resilient to burning, at least when mature. Paperbarks *Melaleuca* species, Cypress-pines *Callitris intratropica*, Cooktown Ironwood *Erythrophleum chlorostachys* and a few other tree species also dominate some savanna areas. The density of tree cover varies widely through the savannas from little or none on regularly flooded areas, to denser forests on some of the higher rainfall areas, such as on the Tiwi Islands and the west coast of Cape York Peninsula.

FIGURE 2.3 **OCCURRENCE OF CLIFFS & TOPOGRAPHIC RELIEF IN AUSTRALIA**

Note that relief is defined as the difference in elevation between divides and valley bottoms for sample circular areas of 25 km².

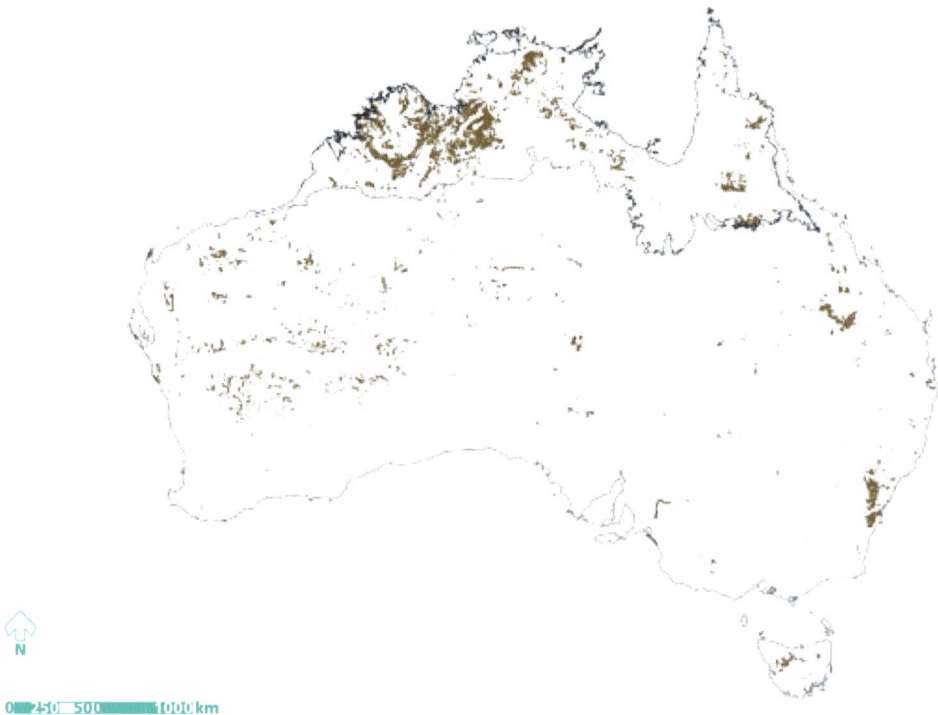

Source: Maps derived from all cliff lines shown on 1:100,000 topographic maps. Data collated by John Stein.

FIGURE 2.4 **BROAD VEGETATION TYPES OF NORTHERN AUSTRALIA**

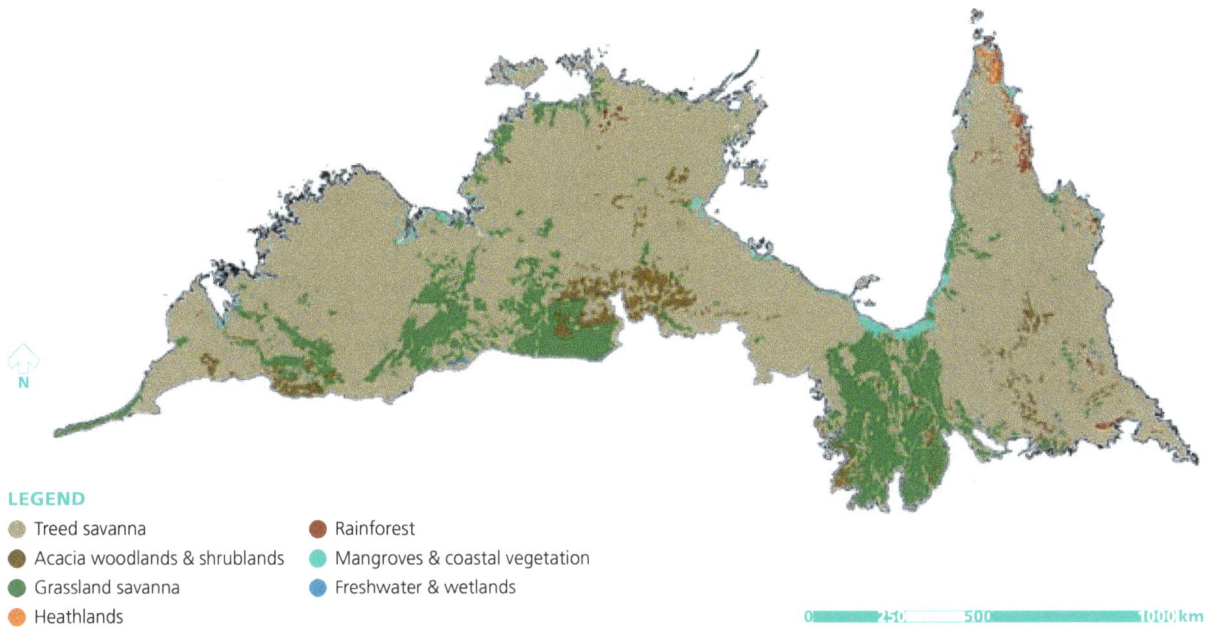

LEGEND

- Treed savanna
- Acacia woodlands & shrublands
- Grassland savanna
- Heathlands
- Rainforest
- Mangroves & coastal vegetation
- Freshwater & wetlands

0   250   500   1000 km

Rainforests

In areas where surface or near surface water is available during the Dry season, and on unusual soils, or in locations that provide protection from fire, the development of non-savanna vegetation types occurs.

Rainforests cover around one million hectares of the wet-dry Northern Australia, including about 7000 hectares in the Kimberley and 268,000 hectares in the Northern Territory.[1] This is less than 1% of the land area, but these rich and fragmented forests support a disproportionate amount of the biological diversity of the North. Various types of rainforest are recognised in the North, including littoral rainforest (on coasts), vine thickets or 'dry' rainforest on rocky outcrops, and riparian or gallery forest along rivers. The largest contiguous area of well-developed tropical rainforest occurs in the McIlwraith Range–Iron Range area on the east coast of Cape York Peninsula. There, the coastal ranges induce more rainfall and cloud drip during the Dry season, conducive to the development of extensive rainforest areas. Elsewhere in the North, rainforest is restricted to tens of thousands of small and discontinuous patches, mostly smaller than 200 hectares in size (McKenzie *et al.* 1991; Russell-Smith *et al.* 1992). These patches are usually in moister areas

WHAT IS A TROPICAL SAVANNA?

Tropical savannas are found, or were found, in parts of Australia, Africa, South and Central America, India and South-East Asia.

Tropical savannas comprise a distinctive landscape typically with a tall dense grass layer with or without trees, that occurs throughout the world's tropics. Tropical savannas can be almost treeless grasslands or denser woodlands ('tropical savanna woodland') – as long as the canopy cover of the trees is not so dense that it shades out the grass.

such as at springs or along perennial rivers, or in areas protected from regular fires, such as gorges and rocky slopes in sandstone country.

Mangroves

Mangroves – the forests of the sea – cover approximately 1.5 million hectares of Northern Australia. Small areas of mangrove ecosystems are found on nearly all parts of the northern coast. In a smaller number of areas in protected estuaries and bays, large and complex forests have formed with several mangrove tree species. Apart from fishers, mangroves are often overlooked by residents and visitors to the North because they tend to have

---

1   In the Wet Tropics around Cairns are another 835,000 hectares of rainforest.

TERMITES OF THE TROPICAL SAVANNAS

Northern Australia is a big country shaped by a small insect – the termite. In many places the very look of northern savannas owes much to the mounds built by this social insect. Indeed these landscapes have given rise to one of the most diverse range of termite mounds in the world: from the enormous buttressed 'cathedrals' of spinifex termites, to the remarkably aligned 'magnetic' mounds and miniature cities of columns built by various *Amitermes* species. Even more termite species – around three quarters of those found in Northern Australia – are hidden from view, building nests within trees or underground. As far as we can tell, most of these species are endemic to Australia.

In part, the termites' remarkable success in the seasonally-dry tropics is due to this mechanism of surviving harsh, dry conditions in the humid micro-climates of their nests, a luxury afforded to few other animals. Given that these nests appear to provide stable internal conditions for termites in a wide range of environments, it is perhaps not surprising that many nests appear to be shaped by those different environments. This is seen in *Amitermes laurensis:* in well-drained habitats it builds small dome-shaped mounds, yet in seasonally flooded flats it constructs huge mounds aligned along a north-south axis, often hundreds of times larger. Studies on this species and *A. meridionalis,* which also builds oriented mounds, suggest that such mounds are an adaptation to the seasonally waterlogged conditions: the high surface-area shape oriented north-south creates a stable environment for living above the ground in flooded habitats where migrating to an underground refuge is impossible (Jacklyn 1992).

Termites can be extremely abundant in certain areas. Savanna habitat covered in the large nests of spinifex termites, *Nasutitermes triodiae,* may support a few hundred tonnes of termite per square kilometre – far greater than the typical biomass of cattle in a similar-sized area. This abundance is no doubt partly due to the fact that termites consume a widely available resource most other animals cannot exploit: cellulose. By virtue of symbiotic bacteria or protozoa in their gut, termites can digest the cellulose and lignin present in a wide range of wood and grass, living or dead, as well as the plant material in litter and soil – with individual species specialising on particular sources of cellulose.

The underground activity of termites assists in recycling nutrients otherwise locked up in wood and dead grass. Such insect-driven nutrient cycling may be more important in Australia than it is in other tropical savannas like those of Africa, where large herbivores are more significant (Andersen and Lonsdale 1990).

Peter Jacklyn

dense populations of both mosquitoes and saltwater crocodiles. However, these forests are exceptionally productive, are vital nurseries for fish, and support many specialised species that live only in mangroves.

Salt-marsh flats

On coastlines with little elevation and low rainfall, extensive salt-marsh flats occur behind a coastal band of mangroves. These distinctive, wide, white and brown flats of salt and dried mud receive only occasional tides. Because they are not regularly flushed, salt accumulates and the areas are too saline for even mangroves to survive. Some highly tolerant shrubs and herbs can survive on some parts of the flats. These ecosystems form spectacular landscapes when seen from the air in places such as the estuaries of the southern Gulf Country coast and Cambridge Gulf in the Kimberley.

Tropical heathlands

Tropical heathlands replace savannas in a few districts on substrates with poor water-holding capacity and low fertility. These low shrublands have a high diversity of plant species within small areas. This vegetation type is most extensive and diverse in some coastal areas in the north-east of Cape York Peninsula and in some rocky escarpment areas, such as in the stone country of Arnhem Land, where heathlands occur on skeletal or infertile soils.

Freshwaters

Between Broome and Cooktown are more than 60 major rivers and hundreds of smaller streams flowing directly into the sea. Combined, these rivers and their tributaries extend over one million kilometres and carry nearly two-thirds of Australia's freshwater.

River flows in the North are extremely variable. Flows vary within a year, between the Wet and Dry seasons, and between different years. This is ecologically important, because floods, including extreme floods, drive the natural productivity of many wetland, estuarine and marine systems.

Most rivers, even the larger ones, are ephemeral in most years, shrinking to non-flowing pools in the Dry. Aquifer-fed streams such as the Daly River in the Top End, the Gregory River in the Gulf, and the Jardine River on Cape York Peninsula,

continue to have significant flows even at the end of the Dry season. Such aquifer-fed perennial rivers are especially important for many terrestrial and aquatic species.

Extensive floodplains in the near-coastal lowlands adjacent to some of the largest northern rivers form some of Australia's largest and most diverse wetlands. These include permanent or semi-permanent billabongs and waterholes along major rivers, and extensive ephemeral wetlands that fill seasonally or during major floods. Examples include the Kakadu wetlands (recognised as of international significance through Ramsar and World Heritage listing) and the Southern Gulf Aggregation. The latter is Australia's largest wetland area and forms when several southern Gulf Country rivers merge in major floods to create a single, vast wetland of around two million hectares.

BIODIVERSITY

Northern Australia supports an abundance of plant and animal life. Many are endemic, occurring nowhere else in the world. Many form spectacular features, such as the dense aggregations of Magpie Geese and other waterfowl that congregate on the northern floodplains. Many are characteristic icons such as the boabs of the Kimberley and eastern Top End, or the saltwater crocodiles that inhabit many northern coasts and waterways.

Much of Northern Australia is recognised as being of outstanding national or international significance for biodiversity. For example, the North Kimberley and Einasleigh and Desert Uplands (in north-central Queensland) are listed among the nation's 15 recently recognised biodiversity hotspots, and Cape York Peninsula and the sandstone plateau of western Arnhem

BOABS & BAOBABS

The huge, gouty, swollen trunk capped with a wide, branching canopy is a distinctive feature of baobabs. The single Australian species *Adansonia gibbosa* (formerly *A. gregorii*) is known throughout the Kimberley of WA and the Victoria River district of the Northern Territory by the contraction to Boab. Although sometimes erroneously termed bottle-trees, they are not related to the bottle-trees (*Brachychiton* species) of inland north-eastern Australia. Madagascar, with six species of baobabs, is the centre of diversity while but a single species occurs throughout a vast area of Africa. Given that baobabs are highly valued sources of shelter, food, fibre and medicines by Indigenous people, some argue that their extensive distribution has been due to human activity. This common usage has led to much speculation on the origins of the single Australian species.

Could the Baobab have been brought to Australia by the first people out of Africa who skirted the then low sea-level shorelines around the Indian Ocean some 50–60,000 years ago? Or, much later, introduction by Austronesian voyages some 1500 years ago, or later still by Arab traders? Or, could some of the large seed pods have floated across from either Africa or Madagascar, like the giant eggs of the now extinct elephant birds of Madagascar, that have been washed up on the beaches of south-western Australia? Or, are the baobabs yet another example of divergence from a common origin after the break up of the ancient Gondwanan super-continent?

Earlier studies of their physical features and floral structures indicated that the closest relationship of our Australian baobab was with a subgroup of the Madagascan species. However, a more recent and very comprehensive set of genetic analyses places the Australian species closest to the single African species (Baum *et al.* 1998). This close genetic relationship effectively rules out the ancient Gondwanan connection and divergent evolution. But, until more detailed genetic analyses can tell us how long ago the African and Australian species diverged, the jury is out on a possible human introduction.

Early explorers noted Aboriginal usage of the boab's fruits, seeds, bark and pith, and the correlation between good camp sites and boab distribution. This should not be interpreted as evidence of direct cultivation, but simply as a combination of discarded fruit and seeds together with favourable sites in valleys for boab establishment. The absence of boabs from favourable sites and climate much further east, where they flourish when introduced, remains a puzzle.

Baobabs are at the centre of myths and legends in Africa and Madagascar and are widely believed to live for thousands of years. So far, the oldest baobab dated was around one thousand years old. In Australia, very old baobabs may be surrounded by a mini-forest of their own seedlings. Old trees resist fire, but young trees may be killed by fire (Bowman 1997). Long-term survival of boabs in north-western Australia will depend on appropriate fire management strategies.

Henry Nix

Land are even richer (Abrahams *et al.* 1995; Mackey *et al.* 2001; Woinarski *et al.* 2006a).

The plants of Northern Australia comprise a hybrid assemblage, including not only many characteristic Australian endemic families, but also some groups whose distribution is centred mostly in the tropics of other continents (Bowman *et al.* 1988). Many species are also shared with New Guinea, particularly so for species found on Cape York Peninsula, in rainforests and in savannas (Nix and Kalma 1972). Other coastal species are shared with Indonesia, or more widely across tropical coastal areas (Woinarski *et al.* 2000a). Ignoring the actual species composition, the broad vegetation structure in Northern Australia may appear almost identical to climatically similar areas of Africa, India or South America.

Many plant species occur widely and commonly across Northern Australia (Clarkson and Kenneally 1988). But others are very rare and/or narrowly restricted. Some parts of Northern Australia have notably high species richness or endemism. The Kimberley supports about 2000 species of native plants (Wheeler *et al.* 1992), of which about 300 are endemic (Beard *et al.* 2000); Cape York Peninsula supports about 3000 species of native plants including at least 260 endemic species (Abrahams *et al.* 1995; Neldner and Clarkson 1995); and the appreciably smaller (~30,000 km²) plateau of western Arnhem Land includes at least 170 endemic plant species (Woinarski *et al.* 2006a).

Native invertebrates (such as insects, spiders and centipedes) are relatively little studied in Australia compared with other animal groups, and particularly so in Northern Australia. Thousands of species remain scientifically undescribed and unnamed. For most species little or nothing is known of their habitat, distribution or ecology. However, everywhere on land they constitute the great bulk of species diversity. In Northern Australia, two insect groups, ants and termites, stand out as being abundant and of particular ecological significance, and are showcased in the accompanying text boxes. By national standards, Northern Australia also supports a high diversity of many other invertebrate groups, including moths and butterflies (where the tally from Northern Australia comprises more than 65% of all Australian species), and grasshoppers. For example, Kakadu

DARWIN WOOLLYBUTT

Eucalypts are the predominant tree species in Northern Australia's savannas, and the Darwin Woollybutt *Eucalyptus miniata* is one of the most common and widespread, occurring from the Kimberley to Burketown in Queensland. It occurs in monsoonal areas with annual rainfall of 600–1500 mm. Its preference for the well-drained sandy or sandy loam soils that are widespread across Northern Australia allows this species to extend across a large swathe of the savannas. Darwin Woollybutt is an evergreen tree, which typically grows to about 15–25 m in height. The limbs and upper trunk are smooth and white but the lower trunk has fibrous bark; and it is this characteristic that gives this species its common name.

Like many savanna species, the Woollybutt's reproductive cycle is closely attuned to seasonal patterns, commencing with the development of flower buds in the early Dry season and finishing with seed released from gumnuts at the onset of the Wet season rains. This cycle, completed in only nine months, contrasts with eucalypts from temperate Australia, which take several years to go from flowering to seeding. Between April and July, the Woollybutts provide a vibrant flowering display with large, bright orange flowers throughout the canopy. The massive gum nuts (up to 6x5 cm) contain one of the largest seeds of all the eucalypts. The nectar-rich flowers and large gum nut seeds offer valuable food for birds, bats and insects, such as the native bees. Black Cockatoos are commonly seen using their strong beaks to strip gum nuts from trees and gouge out the seed.

Most eucalypts in southern Australia require heat from fires to open their seed capsules, resulting in a mass release of seeds. By contrast, the seeds of the Darwin Woollybutt are released from the gum nuts as soon as they ripen, so each year there are a small number of seeds released and no seeds are stored in the canopy from year to year. The seeds that fall to the ground either germinate soon after the early Wet season rains, or decay – so there are no seeds stored in the soil from year to year. The seedlings quickly develop a lignotuber and a deep root system that helps them survive the Dry season and provides some ability to regenerate after fire. Despite this survival mechanism, new seedlings are rare because the period of flowering and seed production coincides with frequent fire. Even the relatively cool fires of the early Dry season can kill flowers and the developing seeds, and hot fires can also reduce the amount of flowering in the following years. Seedlings that are burnt in their first year are unlikely to survive, and a fire-free interval of two or three years is probably needed for seedling establishment.

Flower of Darwin Woollybutt. *Photo by Glenn Walker*

Woollybutt seedlings that do survive to about half a metre are highly resistant to fire or other stresses due to their ability to resprout from a well-developed lignotuber. Frequent fires in the savannas maintain the seedlings in the suppressed woody sprout layer. It is not known how long a fire-free period is required for the sprout to develop into a tree. The high level of fire resistance is also a characteristic of the adult trees and tree survival is high even in sites with annual fire, although survival does decrease with hotter fires. Fire damage and death is often the result of fire entering the trunk or limbs which have been hollowed by termites. One study in Kakadu showed that more than 80% of Woollybutt trees had hollow trunks. Although the hollows contribute to the vulnerability of the individual to fire effects, they provide nest sites for savanna animals and the hollow limbs are highly valued by didgeridoo makers. These values add to the importance of this ubiquitous and marvellous inhabitant of our tropical savannas.

SA Setterfield

National Park alone is known to support 161 grasshopper species (Andersen *et al.* 2000).

The vertebrate species – birds, mammals, reptiles, frogs and freshwater fish – of Northern Australia are diverse, often highly abundant, and include many spectacular species. The distribution of individual species largely mirrors the distribution of vegetation. Most species that use savannas (e.g. Frill-necked Lizard, Blue-winged Kookaburra, Black Kite, Black Flying-Fox) are widespread across the North. Animal species that depend on rainforests, heathlands and rocky escarpments typically have more restricted ranges in the scattered patches of these habitats. In these habitats many localised species have evolved due to their long separation and isolation from other areas of similar habitat. This contrasts with most species using savanna which have been able to find connected habitats right across the North. The rainforests of the east coast of Cape York Peninsula are particularly rich in such endemic species; the magnificent Riflebird,

Palm Cockatoo, Eclectus Parrot and Green Python are examples. Rocky environments also support a distinctive fauna, including Rock Pigeons, Rock Wallabies, Rock Possums and Rock Rats). Mangroves likewise support a distinctive fauna, including species such as the Red-headed Honeyeater and Mangrove Monitor.

Total species richness in Northern Australia is high for most vertebrate groups. About 460 bird species (appreciably more than half of the Australian tally) have been recorded in Northern Australia; along with about 110 native mammal species (about one-third of the Australian tally) and approximately 40% of Australia's reptiles. About 225 freshwater fish (more than 75% of the Australian tally) are known in the North. Indeed, more native fish species may be found in one waterhole in Kakadu than are known in the entire Murray-Darling system.

SO MANY ANTS!

Ants are the dominant faunal group in terms of biomass, energy flow and ecosystem function throughout Australia's tropical savanna landscapes. They contribute up to 30% of the total biomass of the savanna fauna, and play key roles in soil structure, nutrient cycling, invertebrate biodiversity, seed dynamics and plant growth.

Ant diversity, productivity and behavioural dominance in Australian savannas are exceptional by world standards. The total savanna fauna comprises about 1500 species (Andersen 2000), with more than 100 species routinely occurring within less than one hectare (Andersen 1992). These are among the most diverse ant communities on Earth.

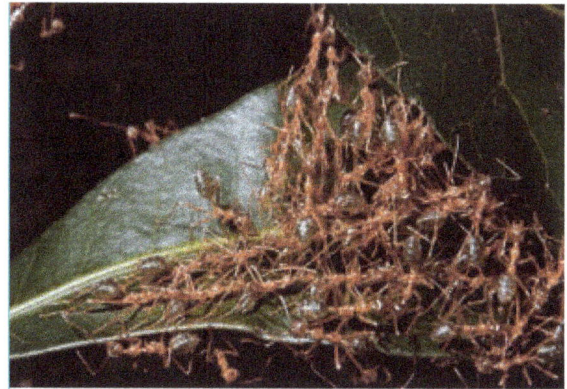

Green tree ants are found across the North. *Photo by Kerry Trapnell*

Savanna ant communities have a characteristic functional group signature, similar to that occurring throughout arid Australia (Andersen 1993a). Typically, communities are dominated numerically and functionally by highly active and aggressive species of *Iridomyrmex*, led by the familiar meat ants *I. sangunineus* and *I. reburrus* (Andersen and Patel 1994). Communities also feature a rich array of taxa whose high degree of morphological, behavioural, physiological or chemical specialisation are indicative of a long evolutionary history of association with behaviourally dominant ants. For example, species of the highly thermophilic genus *Melophorus* (furnace ants) restrict their foraging to the hottest part of the day, at which time they have exclusive access to food resources. The shield ants of the genus *Meranoplus* have highly developed protective morphology accompanied by remarkable defensive behaviour: when harassed they retract their antennae into deep grooves in their head, tuck their legs under a dorsal shield, and lie motionless in a foetal position. There is also a wide range of specialist seed harvester ants, with up to 20 occurring at a single site, each specialising on a different set of seeds (Andersen *et al.* 2000a).

Species of *Iridomyrmex* and their associated highly specialised taxa become less abundant with increasing vegetation cover, and are absent altogether from patches of rainforest. The most common ants in rainforest are broadly adapted and opportunistic species of *Monomorium*, *Pheidole*, *Paratrechina* and *Tetramorium*. However, rainforest ant communities also include a range of specialist forest species, most of which originated in the humid tropics of North Queensland (Reichel and Andersen 1996). The latter include the notorious Green Tree Ant *Oecophylla smaragdina*, a highly aggressive, leaf-nesting species that occurs also throughout the South-East

Asian region. Many other specialist rainforest species are also arboreal, a very uncommon feature in savanna landscapes, where ant activity on vegetation is dominated by ground-nesting species (Majer 1990).

Changes in fire management can cause a rapid shift in ant functional group composition, changing the balance between 'sun-loving' *Iridomyrmex* and associated specialists on one hand, and the generalists of denser vegetation on the other (Andersen 1991). Indeed, there is a continuum from *Iridomyrmex*-dominated systems of annually burnt savanna at one end, to *Oecophylla*-dominated forests in the long-term absence of fire at the other (Andersen *et al.* 2006).

Ant communities are also highly sensitive to disturbances associated with other land management activities, such as grazing and mining (Majer 1984; Andersen 1993b; Woinarski *et al.* 2002; Hoffmann 2000). For example, impacts of emissions from the copper smelter at Mt Isa in north-western Queensland could be detected on ant communities up to 35 kilometres away (Hoffmann *et al.* 2000). Such sensitivity to disturbance, combined with their ecological dominance and ease of sampling, make ants valuable bioindicators for environmental monitoring (Andersen and Majer 2004).

Alan Andersen

1 Painting Up, Banjo
Paterson and Griffith
Patrick Kowanyama, Cape
York Peninsula. *Photo by
Kerry Trapnell*

2 The cattle industry has
driven much of the
economic development of
Northern Australia. *Photo by
Mark Ziembicki*

PEOPLE OF NORTHERN AUSTRALIA

Indigenous ownership and management

With more than 50,000 years of continuous
settlement by Aboriginal peoples, Northern
Australia is the home of more than 130 language
groups. From the Kaanju and Wuthathi on
Cape York Peninsula in the east, to the Bunuba
and Bardi in the west, there are thousands of
clan estates within these language groups.

Clan estates were, and in many places remain,
the foundations of a continent-scale, regionally
distinct, social and economic life. Country
was managed according to specific laws and
customs. This management shaped and created
many of the characteristics of the North that
we see today. Particular vegetation types were
favoured or reduced by management decisions,
such as when and how to burn. Different
species responded in varying ways to fire
regimes, hunting pressures and other practices.

Indigenous people today comprise a major
component of the population and landownership
of Northern Australia, more so than most other
parts of Australia. In many areas, Indigenous
people retain intricate environmental knowledge
and skills in land management. The future
of Northern Australian environments is
inexorably and intricately tied to their future.

Changes in customary land management

With the colonisation of Northern Australia
in the nineteenth century, the patterns of
Indigenous settlement and land management
changed across the North, often very
rapidly. On more fertile and accessible lands,
pastoralism rapidly changed the management
of country, even if the Traditional Owners
remained physically present. Even in areas
that have never been permanently settled by
Europeans, the disrupted cultural, economic
and social situations meant that people
often left their country, or were removed,
and customary practices were altered.

Recent studies suggest that the maintenance
of customary practices may be a key factor in
retaining biodiversity and environmental health
in the North, and the ecological connections
and processes that maintain these values
(e.g. Yibarbuk *et al.* 2000; Bowman *et al.* 2001b).

Changes arising from colonial settlement – many
of which continue today – have had serious
impacts on Indigenous land management
practices. This is compounded by the fact
that there are now fewer people dispersed
across and managing the landscape than at
any previous time in the 50,000 or so years
of human presence in Northern Australia.

Early European and Asian settlement

Since at least the early eighteenth century, there was regular communication between the Indonesian islands and the North, especially in the Top End where the Maccassan people came every year to fish for trepang (sea-cucumber) (Macknight 1976). In the 19th century, European settlers moved into the North. The first serious forays by Europeans into the inland of the North began in the mid-1800s onwards. Settlement was much later than in southern and eastern Australia. Some of the more isolated areas taken up by economic development, such as the central Kimberley, were only first leased to pastoralists in the 20th century. Only the parts of the country most inhospitable to cattle (such as parts of Arnhem Land, Cape York Peninsula and the north Kimberley) were not given over to pastoralists.

The gold rush

The search for two different resources – gold and other minerals, and well-watered grazing lands for sheep and cattle – drove the great bulk of the European invasion and settlement. From the 1870s onwards, a series of gold rushes occurred, notably in Halls Creek in the Kimberley, Pine Creek in the Top End, Croydon in the Gulf Country, and the Palmer River Goldfields and other districts in southern and central Cape York Peninsula. Some of these fields were short lived and were abandoned within a couple of decades, leaving ghost towns, overgrown mine workings and weeds. Other gold fields, such as the Pine Creek district, have remained periodically active to the present day. Through the twentieth century, a broader range of minerals were exploited in fields throughout the North.

Cattle grazing

The first stock were brought to Northern Australia in the isolated and short-lived settlements of the northern coast of what is now the Northern Territory, including Port Essington in 1842. More enduringly, pastoral properties were taken up in central and western Queensland in the 1850s, and pastoralism spread progressively westward over the next few decades. The savannas provided the allure of relatively rapid returns from country where abundant grazing was available without any clearing. However, resistance from Aboriginal landowners, social isolation, poor-quality feed, diseases of people and stock, and long distances

from markets, all combined to make the development of the pastoral industry relatively slow and spasmodic. By 1910, the current extent of pastoral country had largely been reached. Since then, pastoralism has intensified in most regions, depending largely on market prices.

One important change was the introduction of *Bos indicus* cattle (Brahmans and related tropical breeds) in the 1950s. These cattle gradually replaced the *Bos taurus* (European breeds, such as Herefords and shorthorns), which struggled with tropical conditions. This change in stock has led to increased profitability and more intensive grazing in many regions.

Development of the North

While gold, other minerals and the search for pastures drove most of the new settlers, the North has also been punctuated by endeavours to establish major centres of agriculture. Notable examples include attempts to establish sugar cane in the Daly River of the Northern Territory in the 1870s, and the Ord River scheme commencing in the late 1960s. Many of these attempts failed – often at great expense to those involved – due to the harshness and caprice of the monsoonal climate (Bauer 1977; Taylor and Tulloch 1985) and limited areas of fertile soils. In Northern Australia, cropping and horticulture have been established only in small areas on the Ord, the Daly River, around Darwin and in parts of the Gulf Country.

TABLE 2.1 **REGIONAL STATISTICS OF FOUR REGIONS**

Note that for simplicity, we have excluded the areas of central-north Queensland in the study area, such as the Einasleigh Uplands.

	Kimberley[a]	Top End[b]	Gulf[b]	Cape York[b,c]
Area (km²)	300,000	440,000	425,000	137,700
Population (approx. % Indigenous)	34,900 (in 2004) (47% in 2001)	160,000 (30%)	50,000 (25%)	18,000 (60%)
Major towns (population)	• Broome (12,000) • Kununurra (6000) • Derby (5000) • Halls Creek (3600) • Fitzroy Crossing (1150) • Wyndham (1000)	• Darwin (78,000) • Katherine (16,500) • Nhulunby (4000) • Jabiru (1700) • Borroloola (1500)	• Mt. Isa (22,000) • Charters Towers (11,000) • Cloncurry (2500) • Normanton (2500) • Hughenden (1500) • Doomadgee (1200) • Karumba (700)	• Mareeba (6900) • Thursday Island (2500) • Weipa (2200) • Cooktown (1400) • Kowanyama (1200) • Lockhart River (800) • Bamaga (750)
Major economic sectors by gross domestic product (GDP) contribution	• Government services • Defence • Mining • Tourism • Pearling • Cattle • Horticulture	• Government services • Defence • Mining • Tourism • Cattle • Fishing	• Mining • Government services • Tourism • Commercial & recreational fishing • Cattle	• Mining (50% of formal economy) • Government services • Tourism • Cattle • Fishing
Employment by sectors	• Government services • Tourism • Cattle • Mining • Pearling • Horticulture	• Government services • Defence • Tourism • Cattle • Mining	• Government services • Tourism • Mining • Cattle • Commercial fishing	• Community services • Public administration & defence • Mining • Cattle • Fishing
Land tenure	• 20% Aboriginal land • 7% Conservation • <1% Cropping/ Horticulture • 2% Defence • 69% Pastoral lease • <1% Production forestry • <1% Urban/ Peri-urban • 1% Wetlands/Rivers/ Coastal waters	• 39% Aboriginal land • 11% Conservation • <1% Cropping/ Horticulture • 2% Defence • 47% Pastoral lease • <1% Mining • <1% Urban/ Peri-urban • <1% Wetlands/ Rivers/Coastal waters	• 7% Aboriginal land • 6% Conservation • <1% Cropping/Horticulture • <1% Defence • 83% Pastoral lease • <1% Mining • 2% Production forestry • <1% Urban/Peri-urban • <1% Wetlands/Rivers/Coastal waters	

Population estimates from latest available Australian Bureau of Statistics figures (www.abs.gov.au)

a Kimberley Appropriate Economy Roundtable Proceedings, Australian Conservation Foundation 2006

b Australian Bureau of Statistics

c Cape York Peninsula Development Association website: www.the-cape.info accessed February 2007

Data on land tenure provided by relevant state and territory government departments of Western Australia, Northern Territory and Queensland. Note that statistics are for current tenure and may not reflect actual current usage, or past usage, such as previous mining tenures.

For many Europeans, the North was a place to chase wealth, and then leave for their real home country in southern Australia or overseas, and relatively few people (mostly pastoralists) sought to establish long-term roots in the North. This resulted in an unsettled European population that was slow to understand how the northern country really worked.

This trend has been exacerbated by persistent myths of the North as an area of great natural resources and wealth; of the North becoming the 'food-bowl of Asia'; of the need to 'populate the North' as a buffer to invasion by our Asian neighbours. Along with these myths come proposals for large 'nation-building' projects such as the Ord River dam. These myths and the big projects they spawn have proved a graveyard for the dreams and money of governments and individuals. Strangely, the myths have usually ignored real development possibilities, such as the incremental improvement of the more profitable tourism and pastoral industries.

REGIONS OF THE NORTH

While there are obvious similarities in environments, landform and climate across the North, there are also some notable regional differences, in part arising from political segmentation and development history.

The North can be usefully divided into four regions – the Kimberley, Top End, Gulf Country and Cape York Peninsula (Figure 1.1, Chapter 1). Each region is briefly described below, and Table 2.1 gives some of the vital statistics for the areas.

The Kimberley

Rugged ranges and escarpments dominate the northern and central Kimberley. Here flat-topped mesas, vertical cliffs and rocky screes provide distinctive habitats, with fire refuges for rainforests and protected overhangs for rock art. In the southern and south-western inland areas,

IRONWOOD

Eucalypts give the Northern Australian woodlands and open forests their dominant species and much of their ecological character. But there are foreign elements amongst these distinctively Australian trees. Perhaps the most notable of these is Cooktown Ironwood. Although this is an endemic Australian tree, its affinities lie elsewhere: the nine other species in the genus

The Ironwood is a fire sensitive tree found across the North. *Photo by John Woinarski*

Erythrophleum are restricted to tropical regions of Africa, Madagascar, continental Asia and Malesia. It provides an example of the ecological melting pot that is Northern Australia.

Ironwood *Erythrophleum chlorostachys* is widespread across Northern Australia: indeed one of its most pervasive trees. European settlers marked it early: its foliage was poisonous to stock, but its wood was hard (one of the densest timbers of any Australian tree), termite-resistant and durable, and much sought during the early decades of European settlement. Around more intensively settled areas, timber harvesters quickly exhausted the resource of large ironwood trees.

Although Ironwood is common and widespread, there are few large trees. It brings unusual ecological traits to this highly dynamic environment: it is long-lived, slow-growing and fire-sensitive. It may take 300–500 years for ironwoods to become large trees (Cook *et al.* 2005). In the current regime of frequent fires, there is little chance of ironwood saplings reaching maturity; and even large trees are now vulnerable to the increasingly intense fires. Ironwood survives in these landscapes largely in a sub-optimal, waiting state. Although it can reproduce by seed, it can also reproduce vegetatively (from root-suckers and lignotubers), and it may stay for many decades in a persistent cycle of growth, fire, loss of stem and leaves, then regeneration from root-stock. Unfortunately, we don't yet know for how long ironwood can maintain this enforced immaturity: the cycles are unlikely to be endless. Where fires are frequent, ironwood may be evident only as a low carpet of suppressed suckers. Large ironwood trees may be present only in the diminishingly small areas in which fire is absent, infrequent and/or mild. Given the opportunity of substantially reduced fire frequency and/or intensity, ironwood could become the dominant tree across much of Northern Australia. Such an opportunity is unlikely to materialise under current management regimes; and the problem for ironwood, and other (typically less conspicuous) species that share its ecological characteristics, is that the unburnt or infrequently burnt refugia are becoming sparser, and there is no respite from the fires across the rest of the landscape.

John Woinarski

the Kimberley meets the Great Sandy Desert and, at these margins, the country is flatter, more arid, and dominated by low woodlands of eucalypts, pindan wattles and grasslands.

The Kimberley is sparsely settled with less than 35,000 people occupying 300,000 square kilometres and, as elsewhere in the North, most people live in the few towns. Large areas in the North and Central Kimberley remain highly isolated from towns and roads. They have very few residents and are rarely visited.

The Top End

The Top End refers to the part of the Northern Territory north of about 18°S – the large square block extending northwards in the centre of Northern Australia. The region has an exceptional diversity of country. The dominant landform are plains and low hills with vast areas of savanna of eucalypts and grass. There are also extensive grasslands in the Victoria River District and on coastal floodplains, mangrove forests such as those around Darwin Harbour, numerous rainforest patches, and the spectacular 'stone country' (the plateau and gorge country such as in Kakadu, Litchfield National Park and western Arnhem Land).

Largely because of Darwin (78,000 people), the Top End has the highest population of any of the four big Northern regions. There is now a small but growing intensively settled area around Darwin, but elsewhere populations are again largely based on towns and mining camps, with a sparse population on country at cattle stations and on Aboriginal homeland outstations. Large parts of Arnhem Land and the south-west of the region remain very isolated, with few residents.

The Gulf Country

The dominating feature of the Gulf Country is the great plain that sweeps around the southern part of the Gulf of Carpentaria. This huge alluvial plain is bordered by the Mt Isa Uplands and the Gulf Fall Uplands and the Barkly Tableland in the west, and by the Einasleigh Uplands to the east. During big Wet season floods, lands bordering the southern Gulf coast are inundated and form the largest wetlands in Australia.

Much of the coastal plain is good quality cattle country and is managed closely for beef production. The highly mineralised belt of the Mt Isa Uplands has produced the long

FIGURE 2.5 **MAP OF LAND TENURES ACROSS THE NORTH**

Note that current tenure does not always correspond with actual current use or past use. For example some pastoral leases may not be actively managed for grazing and many historical mining areas are not shown on the map.

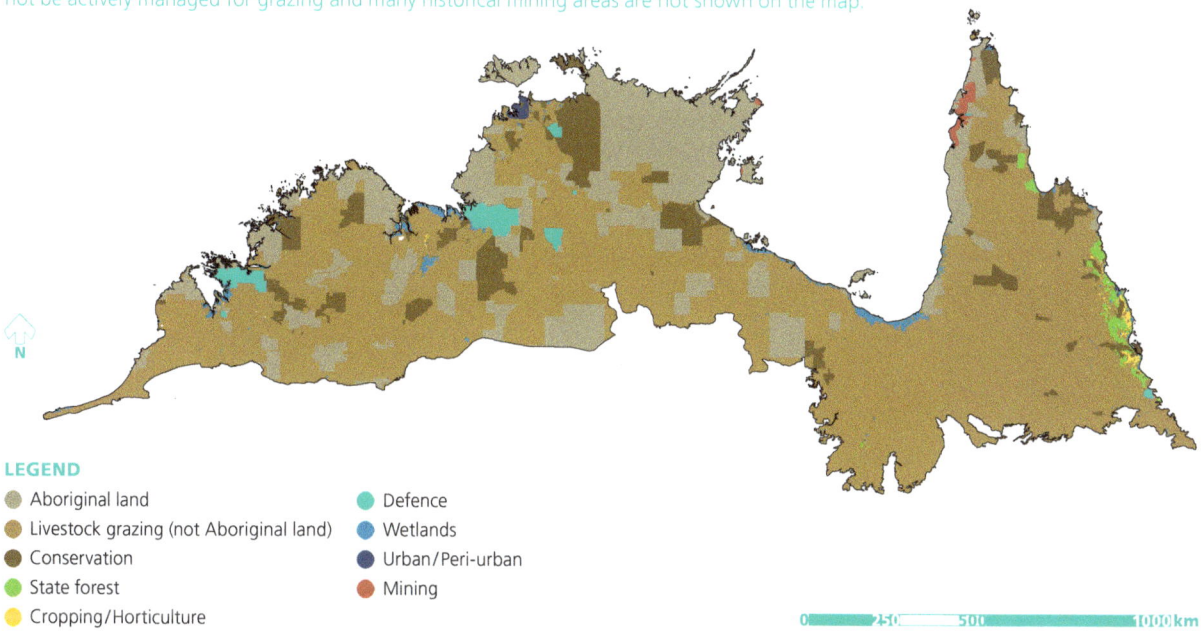

LEGEND

- Aboriginal land
- Livestock grazing (not Aboriginal land)
- Conservation
- State forest
- Cropping/Horticulture
- Defence
- Wetlands
- Urban/Peri-urban
- Mining

0   250   500   1000 km

Source: Prepared by Jane Edwards, databases from National Land and Water Resources Audit.

productive mines and small city of Mt Isa, as well as other mines in the mineralised belt that extends north-west into the Northern Territory.

Cape York Peninsula

The Cape has an exceptional diversity of country. In the east the Great Dividing Range form a mosaic of hill country with rainforest, savanna, mangroves and heathlands. In the west are vast low-lying savannas and seasonal wetlands, an extension of the coastal plain of the Gulf Country.

The Cape is one of the most sparsely populated large regions on Earth. Its small population is concentrated in a few towns, including the mine and mining town at Weipa, which is the dominant industry in dollar terms on the Cape.

CURRENT LAND USE AND ECONOMIES

Land tenure and use

The dominant land tenures in Northern Australia are pastoral freehold and leasehold lands (comprising about 70% of the total area) and Aboriginal-owned lands (about 20%). Smaller areas are devoted to conservation reserves (6%),

military training (1%) and other uses. These figures can be interpreted variably, as some lands have mixed uses, such as Aboriginal lands jointly managed for conservation, or Aboriginal lands run as pastoral enterprises.

The pastoral and Indigenous lands occur throughout most parts of the North. Conservation reserves are scattered across all regions, though are largely absent from the southern Gulf Region in Queensland and the west Kimberley. The small amounts of non-Indigenous freehold land are mostly associated with towns or agricultural areas near Darwin and Kununurra.

Not shown in Figure 2.5 are operational mines. Currently, mines and mining exploration takes up small areas in distinct mining provinces across the North – mostly in the scattered diamond fields of the Kimberley, uranium and gold around Pine Creek and Jabiru in the Northern Territory, bauxite strip mining in eastern Arnhem Land and the west coast of Cape York Peninsula, strip mining for manganese on Groote Eylandt, and a range of metals, phosphate and diamonds from the large north-west mining province in Queensland that extends from Mt Isa into the Gulf hinterland of the Northern Territory.

① View from Gunlom Falls, Kakadu National Park, Top End. *Photo by Glenn Walker*

② Kids cooling off, Kakadu National Park, Top End. *Photo by Glenn Walker*

Economies

The economy of the North is dominated by sectors that depend on natural resources and the provision of government services. However, the main economic drivers of Northern Australia do not mirror land tenure. Precise figures are hard to obtain because the North covers three state jurisdictions; however, government services (including health, education, defence and quarantine) are the largest sectors of economic activity. Importantly, these services involve a continuing and significant transfer of funds by governments from southern Australia to northern regions.

The largest industry sectors in the North are:
- Mining (with sales of minerals worth $2464 million from the Northern Territory in 2004–05);
- Tourism ($1161 million in the Northern Territory in 2003–04); and
- Cattle (more than $1000 million from Northern Australia and central Queensland in 2003).

Relative job numbers do not closely mirror the size of these sectors (Figure 2.6). Government service industries, tourism and pastoralism provide more employment than mining. This composition of different sectors is also very different from the rest of Australia. In Northern Australia there is a lower proportion of jobs in manufacturing and financial services, but a higher proportion in government services and defence, mining, agriculture and fishing.

In addition, many Aboriginal communities have hybrid economies that provide income from a range of government and non-government sources, including sales of artwork, employment and training schemes, provision of services such as quarantine and land-management work, seasonal work on cattle stations, and collection of food and other needs from the bush and sea.

Distribution of people across the North has changed dramatically since European settlement. In 1850, Indigenous people and the first European settlers were highly dispersed across the country. Since then, people have become more concentrated in population centres.

Currently, there are around 250,000 people in the Kimberley, Top End, Gulf Country, Cape York Peninsula and central-north Queensland outside the Wet Tropics. The great majority of people live in towns; others live in smaller towns or mining

settlements; and relatively few people live and work in the country. The sparsity of people in the broader landscape represents a major problem for the North, for example in providing for active management of fire, weeds and feral animals.

CONCLUSION

This chapter provided a brief introduction to the landscape, history and economy of Northern Australia. It is a distinctive land, marked by a climate more related to tropical areas on other continents than to the rest of Australia. Its environments include an idiosyncratic mix of typically Australian and tropical elements, with a distinctive legacy of long-term connections with New Guinea. Its current social and economic features are unlike those of most of the rest of Australia.

In the following chapters we delve more deeply into some of these features. We draw out the manner in which the ecology of Northern Australia works, its conservation values and management challenges. The following chapter describes in more detail the ecological underpinning of the North – how the land operates to sustain its biodiversity and natural resources.

FIGURE 2.6 **EMPLOYMENT BY SECTOR**

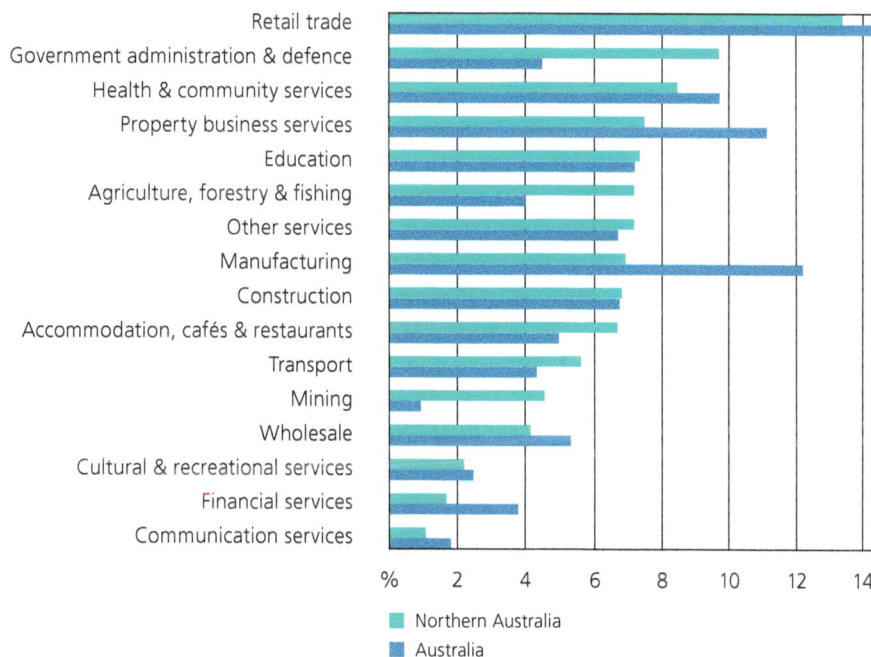

Source: ABARE, Northern Australia Regional Outlook paper, November 2006, www.abare.gov.au/publications.

HOW THE LANDSCAPE WORKS

Most analyses of the conservation values of a region simply document the natural values – catalogues of species, of particular habitats, of beautiful landscapes. We have chosen to look deeper than this simple stock-taking, and seek instead to describe how it is that those assets exist, what keeps them functioning, and what they signify.

This chapter examines the fundamental ecological processes and connections that shape, drive and support the ecosystems and species that make up the environment of Northern Australia.

ECOLOGICAL FUNCTIONING AND LINKAGES

The defining feature of Northern Australia is its pervasive naturalness. Modified lands – scattered towns and relatively small areas of intensive agricultural development – are the exception, influencing but not yet dominating the ecological functioning of the broader landscape.

But how does this landscape work? How do its elements connect? What sustains them? How much can be changed before the

fundamental integrity of the system is lost? If the outstanding values of Northern Australia are to be sustained, these questions need to be answered, and those answers incorporated in land planning, use and management.

These questions are deceptively simple. Ecological science has been far better at understanding the workings of single species in simple fragmented systems than it has in deciphering the complex weave of multiple species operating at varying scales of time and space in large intact systems. Nature is bewilderingly complex, but it is that complexity that gives it its viability, resilience and beauty.

A recent analysis (Soulé *et al.* 2004) identified seven key ecological processes and connections operating at a continental scale in Australia: hydro-ecology; disturbances; long-distance biological movements; strong interspecies interactions; climate change and variability; land-sea connections; and evolutionary processes.

These highly inter-related processes structure the distinctive way the North works as a set of interconnected landscape elements. They connect and drive every aspect of nature.

① (Previous page) Mangrove seedling, Temple Bay, Cape York Peninsula. *Photo by Kerry Trapnell*

② Dry season deciduousness is a characteristic of many plants, including some eucalypts, as shown here. *Photo by Michelle Watson*

③ Flowering following the Wet season rains. *Photo by Atticus Fleming*

If these processes are altered or degraded, then individual pieces of nature change or disappear. These changes may be predictable and immediate, or they may be unpredictable, subtle and delayed. They may be constrained and localised, or the arcane interconnections may mean that repercussions are enacted remote from the source of change.

For example, a change in flows of spring water to a section of a northern river may immediately affect the feeding habitat of Pig-nosed Turtles at the site of the springs. A month later it may have an impact on the

breeding success of barramundi and Magpie Geese tens of kilometres downstream. It may change the success of prawn fisheries off the estuary and influence the fruiting seasonality or success in the coming year of rainforest trees along the river, in turn affecting a colony of Black Fruit-bats currently feeding hundreds of kilometres away at another food source.

Three of the above processes are particularly significant in Northern Australia: *Hydro-ecology (water), Disturbance (fire),* and *Long-distance biological movements.* Highly inter-related, they dominate the distinctive way the North works as a set of interconnected landscapes.

HYDRO-ECOLOGY: THE INTERDEPENDENCE OF WATER AND LANDSCAPE FUNCTION

Water is a pivotal feature in a landscape dominated by long seasonal droughts interspersed with episodes of torrential rain and flooding. Water availability varies dramatically in the landscape over the course of the year, rendering all northern landscapes highly dynamic. From November onwards, north-westerly winds bring monsoonal troughs and cyclones across Northern Australia. By April or May the winds shift and the monsoon rains retreat. For the next seven or eight months average potential evaporation greatly exceeds rainfall and there is a major water deficit. The duration of the Dry season and the location of water during this time become keys to the survival of most species.

Some general features of the hydro-ecology of the North are that:

- The Dry season is long. For many species, it is a time of increasing resource depletion – a time to endure; to shut down; to hunt the landscape for diminishing patches of shelter, food or sustenance; or to take refuge.

- The Wet season is a time of replenishment and revitalisation. The brown lands become green, and the creeks and waterholes fill.

- But the Wet also brings chaos. Destructive cyclones are common, and frequent floods inundate large areas. The lightning storms that herald the Wet season can be rainless and then ignite fierce fires. But amidst the plenty, there can be lack of food for some species such as seed-eating birds; or lack of shelter, such as for floodplain-dwelling rats whose burrow systems may suddenly become inundated.

- The timing and location of the first rains are unreliable. The length of the Wet season, its total rainfall and patterning of rain days are highly variable. This poses great risks for species whose survival through the lean times of the Dry season may be on a knife-edge, or those associated with particular environmental conditions. For example, colonies of Magpie Geese nest only once the floodplains have been inundated to a depth of 30–90 cm, but their nests are then susceptible to further flooding if rains persist. (The variability in the onset, duration and rainfall pattern of the Wet season has also led to the collapse of a range of horticultural enterprises, misled by the simplicity of rainfall averages.)

But, of course, some species prefer the Dry season. Compared with the bleak winters of southern Australia, the balmy days of July in Northern Australia are attractive. Many bird species from southern Australia depend upon yearly northern migrations. And the unpredictability of the Wet season is relative: the North has a far more reliable climatic system than that in most other parts of Australia, where drought may be a frequent and unwanted visitor.

The severity of the Dry season is such that, for many months and over large areas, surface water resources are restricted to a few permanent or semi-permanent water holes and streams, which become key foci in the landscape. Mostly, the supply of water at these locations is not maintained by rainwater, but by water discharging from underground aquifers. These aquifers are in turn replenished during the Wet season.

Where the location of particular soils, aquifers or topographic characteristics permit, moisture availability may be maintained at a site for most or all of the year. These conditions favour water-dependent ecosystems such as monsoonal rainforest patches. Even subtle variation in these characteristics may change the relative advantage of different plant or animal species, providing a landscape characterised by a nuanced patchwork of different habitats. This variety leads to the maintenance of higher levels of biodiversity and more options for species trying to live in the landscape.

④ Many animals, such as the Red-tailed Black Cockatoo, depend on isolated waterholes in the Dry season. *Photo by Lochman Transparencies*

In some regions of Northern Australia, the distribution and interactions of water resources are relatively well-known. Figure 3.1 presents such an example, where the location of springs, water-dependent ecosystems, flow rates of rivers and other hydrological information has been reasonably well-defined. This is a complex system, with broad-scale interconnections between components. Aquifers have a distributional pattern that don't mesh with above-ground landscape divisions; some rivers are far more seasonal than others; and water-dependent ecosystems are highly localised and patchily distributed.

FIGURE 3.1 THE DISTRIBUTION OF SURFACE & GROUNDWATER AT THE END OF THE DRY SEASON IN WEST & EAST ARNHEM LAND

The map shows the minimum volume of water flowing in groundwater fed streams during the Dry season, the location of rainforest patches dependent on groundwater, along with springs and waterholes.

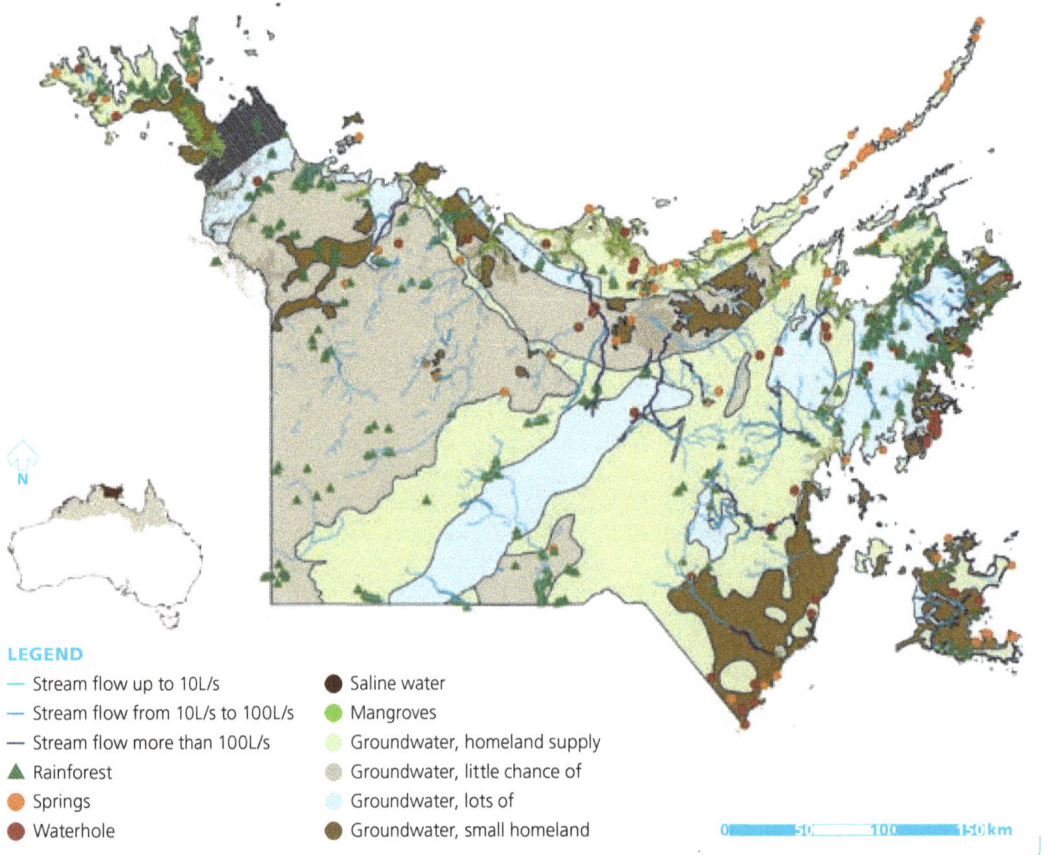

LEGEND

— Stream flow up to 10L/s
— Stream flow from 10L/s to 100L/s
— Stream flow more than 100L/s
▲ Rainforest
● Springs
● Waterhole

● Saline water
● Mangroves
● Groundwater, homeland supply
● Groundwater, little chance of
● Groundwater, lots of
● Groundwater, small homeland

Source: Ursula Zaar, Natural Resources, Environment and the Arts (NRETA).

The characteristics of water availability are important not only for the spatial positioning of different environments and species across the landscape, but also for the timing of resources that those landscape elements produce. Water availability allows plants to grow, and seeds and fruits to be produced. It prompts and allows breeding or emergence of turtles, fish, frogs, crocodiles, aquatic invertebrates, and waterfowl. These responses may be delicately-poised and intricate: seeds of different grass species will germinate in response to different triggering thresholds of rainfall events; seeds of the same grass species will respond more rapidly to the same rain event if they happen to be positioned in even a slight depression; particular levels of river flow will trigger breeding activity in Pig-nosed Turtles but too much river flow may destroy their clutches. In the same year, spatial variation in rainfall characteristics may allow waterfowl breeding in one catchment but not a neighbouring one.

Water interacts with other key ecological processes. For example, relatively high moisture levels – and consequently relatively green vegetation – in the early Dry season means that fires then are generally relatively patchy, cool and small in scale. In contrast, tinder-dry vegetation in the late Dry season provides conditions for fires that are far less manageable, and characteristically hotter and more extensive. This relationship is two-way: extensively burnt landscapes are more likely to further dry out and lose soil moisture. And the fire history around riparian areas will affect stream temperatures, water quality, the chemical and physical properties of the water, and aquatic plants and animals.

To understand, and hence manage, the northern landscapes, the patterning of water availability also needs to be considered at broader spatial and time scales. For example, global climate change will shift environmental relationships across the north. There has been a marked change in rainfall patterning over the last 30 years, with some regions in Northern Australia experiencing more rain and others less (Figure 3.2). If these changes are indeed the harbinger of climate change then we can expect significant ecological impacts. In response to this change, or perhaps to a combination of factors including increased atmospheric CO_2, or changed fire regimes, there have been substantial directional changes in vegetation patterning in Northern Australia. This includes in some areas expansion of rainforest

GROUNDWATER & PLANT COMMUNITIES

Within a given catchment, variation in local drainage conditions is the most important factor determining the kind and distribution of vegetation communities (CSIRO 1953). At the extremes of topographic position, parts of a landscape are seasonally inundated and support grasslands or grass-reed swamp communities; other locations such as ridges experience excessive run-off and support Deciduous Open Forest. The distribution of most vegetation types can be explained by local drainage conditions: for example, on Cape York Peninsula, *Scleandrium leptocarpus* swamp communities occur on sandy slopes and flats that have restricted drainage and remain wet by seepage; *Pandanus* scrubs often form nearly pure stands at margins of swampy grasslands or along drainage channels in heavy soil areas; and Deciduous Parklands of monsoon forest are associated with some of the larger river systems on alluvial flats (Neldner and Clarkson 1995; Mackey *et al.* 2001).

Of particular interest are the riparian forests that fringe most stream-lines. In tidal waters, riparian forest is dominated by mangroves. Further upstream monsoon rainforest develops, with tall tree species such as kapok, cotton and Leichhardt trees. Monsoon rainforest also occurs sporadically as small patches at the headwaters of spring-fed creeks, on levees or on rocky outcrops. Riparian vegetation ecosystems support unique assemblages of plants and animals, and provide a concentration of natural resources (e.g. water, nutrients) that are otherwise scarce in the broader landscape, at least seasonally (Woinarski *et al.* 2000b). These ecosystems attract not only unique wildlife but also human activity as they represent important recreational, cultural and aesthetic resources.

The depth of the water table in the Dry season may have particular significance for vegetation (Sebastien *et al.* 2005). Plants have evolved differing life history strategies to gain the water they need and persist during the Dry season. For example, Freshwater Mangrove *Barringtonia acutangula* and Silver-leaved Paperbark *Melaleuca argentea* use groundwater almost exclusively and grow along river-banks and lower terraces with shallow water tables, whereas Ghost Gum *Corymbia bella* occurs on levees and mostly uses soil water to survive the Dry season. Other species such as *Cathormium umbel-latum* and Darwin Black Wattle *Acacia auriculiformis* draw upon either groundwater or soil water and can occur across the riparian zone.

Brendan Mackey

and other denser vegetation, and 'vegetation thickening' (Crowley and Garnett 2000; Lewis 2002; Bowman *et al.* 2001a; Fensham and Fairfax 2003; Banfai and Bowman 2006). In parts of north-eastern Australia, where the

FIGURE 3.2 **TREND IN ANNUAL TOTAL RAINFALL** (mm/10yrs)

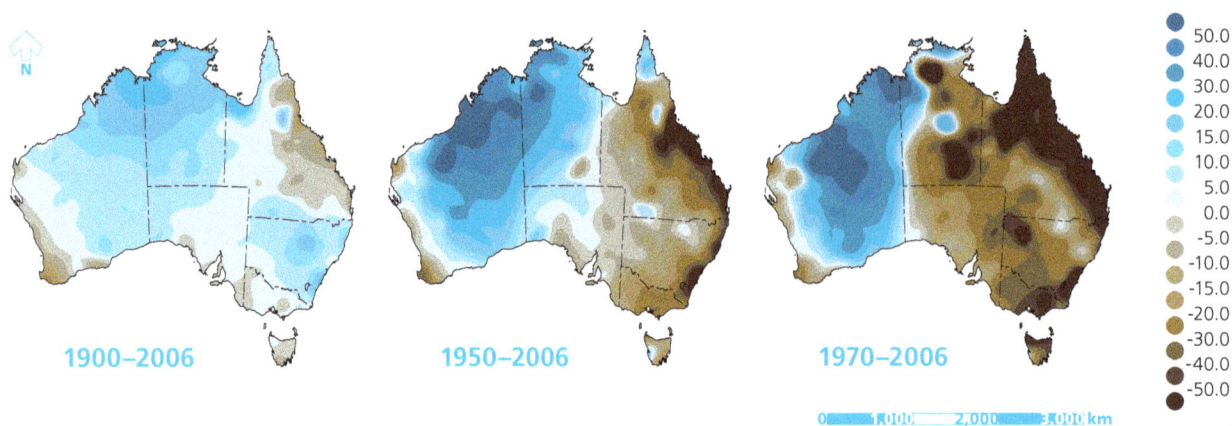

1900–2006   1950–2006   1970–2006

●	50.0
●	40.0
●	30.0
●	20.0
●	15.0
●	10.0
●	5.0
●	0.0
●	-5.0
●	-10.0
●	-15.0
●	-20.0
●	-30.0
●	-40.0
●	-50.0

0   1,000   2,000   3,000 km

Source: Bureau of Meteorology; www.bom.gov.au.

monsoonal rains may be less reliable, tree cover may slowly pulse, increasing in the decades of good seasons and, through extensive tree death, reducing in periods of below-average rainfall (Fensham and Holman 1999).

ECOLOGICAL DISTURBANCE

Ecological systems are dynamic. Their appearance, species composition, vegetation structure and character change over space and time. Much of this fluidity is a direct response to disturbance, a term used in ecology to describe a range of natural or artificial processes and forces acting on a system. The major disturbances operating in Northern Australia are flood, cyclones and fire. Herbivory, particularly by exotic grazing animals, may also meet some definitions of disturbance. However, we consider this issue elsewhere (in Chapters 4–6), and instead restrict consideration here mostly to the ecological role of fire in Northern Australia.

There is apparent unpredictability and inconstancy in ecological disturbance. We may see disturbances as unmanageable and destructive chaos, but they are also forces that can produce differences and diversity at the landscape scale. Of course, some landscape elements experience far more frequent disturbances than others. Riparian systems experience frequent massive flood events which favour plant species that can withstand these disturbances or recolonise rapidly. Conversely, environments in the rugged sandstone ranges have been largely protected from the more mutable and exposed world of

the lowlands over tens of thousands of years. The timeless grandeur people experience when visiting these landscapes reflects the ecological reality that they have sheltered many species that would have little hope in a more frequently disturbed landscape position.

Fire is an inescapable element of the Northern Australian landscape. It is a force of destruction, yet can be the most useful of tools. Ultimately, the prevalence of fire is a consequence of the monsoonal climate. Lands baked dry by the long rainless season invite fire and, even in an unpeopled landscape, would have been regularly and frequently ignited by lightning strike.

For tens of thousands of years Aboriginal people have burnt this landscape. This burning was undertaken in a systematic manner which was deeply considered and embedded within cultural practices. Fires were generally small and patchy, and were lit for ceremony, for hunting, to allow easier travel, and to protect key resources. Poor fire management was deeply censured. Over time, the landscapes and their resources were sculpted by this traditional and consistent use of fire; and we inherited a pattern of species abundance and landscape patchwork that was a product of, and keyed into, this fire regime.

Over the last 100 or so years, that regime has been usurped or broken down. Generally, the current fire regime is far less considered or consistent. Large areas of Northern Australia have become pastoral lands. In smaller and more intensively managed pastoral properties, cattle grazing has greatly reduced grassy fuel loads. Most pastoralists have seen fire as an

LONG-UNBURNT VS BURNT LANDSCAPES

Most of the eucalypt savannas in Northern Australia are burnt every year or two. People living in or visiting Northern Australia will mostly see, and hence typify, such savannas as structurally simple, little more than a tree layer dominated by one or two eucalypt species above a dense layer of tall grasses: there seems to be not much else to the system. But what would this country look like without fire?

There have been several small-scale experiments in Northern Australia that have successfully kept fire out of research plots for periods of more than a decade, and then compared the resulting vegetation features with those of adjacent frequently-burnt sites (Russell-Smith *et al.* 2003; Williams *et al.* 2003; Woinarski *et al.* 2004a). These studies have consistently shown a gradual development of a woody mid-storey layer (typically dominated by plants that produce fleshy fruits), a corresponding decrease in grass cover, and an increase in leaf litter and woody debris. With reduction in fire frequency, some plant species decrease and others prosper. Plants with rainforest affinity may begin to colonise unburnt sites, especially if the sites are near appropriate recruitment areas.

Because these differences in vegetation between frequently and infrequently burnt sites relate to features affecting habitat suitability for many animal species, there are comparable differences in the abundance of many animals between frequently and infrequently burnt sites. Animals that prefer grass and grass seeds are generally more common in frequently burnt sites, whereas those that prefer leaf litter, shade, logs and fruits become more common when sites are unburnt for longer periods. For example, at the site pictured below, the mean abundance of Northern Brown Bandicoot was 1.8 (animals per 100 trapnights) in frequently burnt locations but only 0.3 in long-unburnt locations; whereas that of the Common Brush-tailed Possum was 2.6 in long-unburnt locations but only 0.4 in frequently-burnt location (Woinarski *et al.* 2004a).

The conservation challenge in Northern Australia is to achieve some balance in fire regimes, such that enough of the landscape is frequently burnt (to provide benefit to species such as the bandicoot) and enough of the landscape is infrequently burnt (to provide benefit to species such as the possum). At present, that balance does not exist: in the eucalypt savanna there are few areas that escape fire for more than five years. The possums and their team-mates are losing out.

John Woinarski

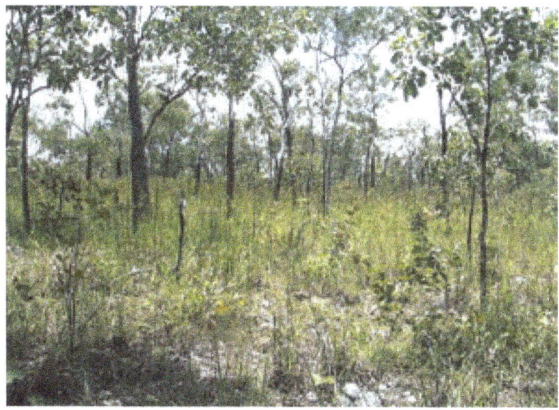

Same study site and landscape: the photo on the right is of a frequently-burnt eucalypt savanna, whereas that on the left is of the same eucalypt savanna unburnt for 26 years. *Photos by John Woinarski*

unwelcome thief of fodder, and consequently fires are typically far less frequent. In contrast, more extensively and particularly on Aboriginal lands, fires are still frequent, but generally their patterning and impact has changed substantially; largely as a result of the landscape being depopulated (Bowman *et al.* 2001b). Consequently, few areas are now managed for fire with the same intricate

care and intimate knowledge as previously (Yibarbuk *et al.* 2001). Most fires now burn far larger areas, are hotter and less controllable.

The current patterning of fire in Northern Australia is most readily apparent from satellite imagery (Figure 3.3). In an Australian context, fire is far more pervasive in the North than elsewhere. Over large areas of Northern Australia, fire is an almost annual occurrence, and only a very small proportion of this landscape has escaped more than 3–5 years without fire.

Typically, fires in Northern Australia are of relatively low intensity consuming the grassy understorey but leaving the tree canopy foliage largely unscathed. However, a feature of fires in Northern Australia is that they can also damage or destroy the largest trees, because these typically are hollowed by termite and fungus at ground level. Where fire can enter the base of these trees, the hollowing acts as a chimney and the fire is taken throughout the trunk, often leading to tree fall and loss of nest hollows and shelter.

The intensity and patchiness of fires varies with season and landscape type. In the early months of the Dry season, the grass moisture content is relatively high, fires typically go out at night, and fires burn unevenly, leaving some patches unburnt within their extent. In contrast, the hotter temperatures and far drier fuels of the later Dry season support far more high-intensity and extensive fires; and fires at this time typically are far less patchy, and may be so intense that they damage the canopy. In many places, lands are burnt in the early Dry season as a deliberate attempt to avoid the perceived risk of more destructive late Dry season fires. Given the limited resources available across the Northern Australian landscapes, such preventative burning is one of the few tools available to land managers.

Much of the flora of Northern Australia tolerates frequent fire. Many of the eucalypts reproduce vegetatively, for dependence upon fire-vulnerable seedlings is too risky a strategy. Many of the understorey plants can withstand frequent

1 A typical savanna fire, Cape York Peninsula. *Photo by Kerry Trapnell.*

FIGURE 3.3 **FIRE FREQUENCY (1997–2002)**

LEGEND

Number of times burnt

- Once
- Twice
- Three times
- Four times
- Five times
- Six times
- Seven times
- Eight times
- Nine times

N

0   250   500   1000 km

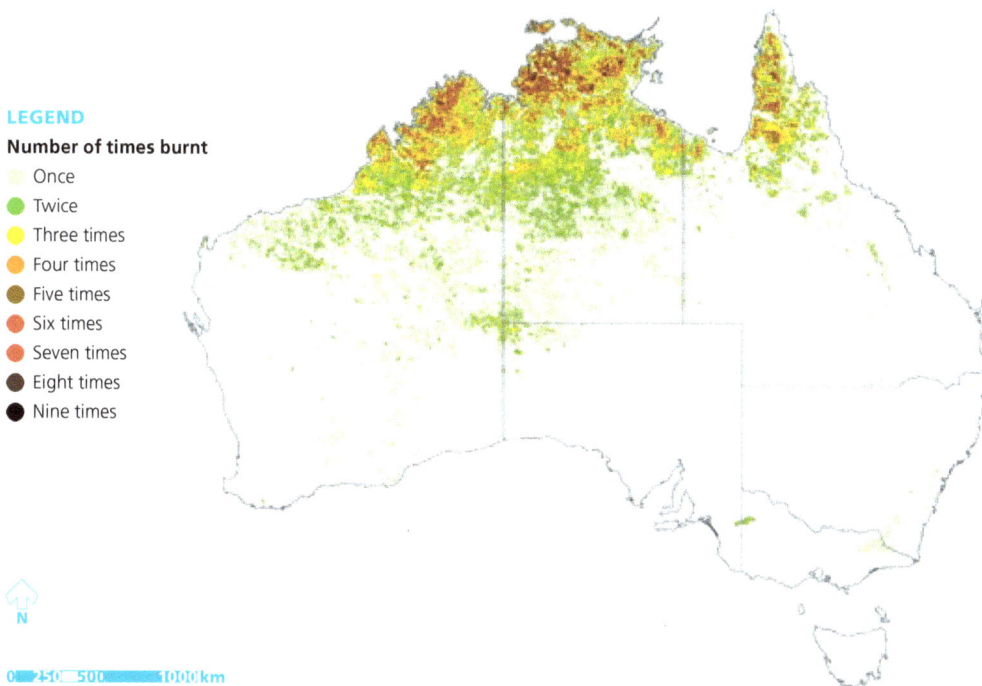

Derived from NOAA-AVHRR satellite image by WA DLI.

defoliation by fire. Other plants have seeds that can withstand the characteristically low intensity fires. However, there is a component of the flora that is adapted to fire regimes characterised by extremely long intervals between very low intensity fires. Such plants are commonly referred to as 'fire sensitive' and include a group of shrub and tree species that reproduce only by seed, and that may take five or more years to reach maturity. Where fires occur more frequently than this maturation cycle, these plant species will inevitably decline and be lost. The Northern Cypress-pine is one such species (Bowman and Panton 1993; Price and Bowman 1994). Many other fire sensitive species occur in sandstone heathlands of Arnhem Land – the plant community that supports the highest diversity of Northern Australia's most restricted endemic plant species (Russell-Smith *et al*. 1998, 2000). Many other fire-sensitive plants are restricted to patches of rainforests, and these may be degraded and lost where exposed to frequent fire. Fire sensitive species are largely restricted to sites that offer some protection from frequent and intense fires, either because they are permanently wet (such as springs) or topographically guarded (such as in gorges and escarpments).

Fire regimes affect the ecology of animal species in many ways. Although fires in Northern Australia are typically of relatively low intensity and slow-moving, there may be some direct casualties in any fire. Animals most at risk include ground or grass-nesting birds (such as emus, quails, Partridge Pigeons, Masked Finches, wrens), slow-moving reptiles (such as Blue-tongued Lizards), and species that shelter in fallen hollow logs (such as some snakes and small mammals), leaf litter (such as spiders and geckoes) and/or dense grass (such as bandicoots). Many vertebrates (including tree frogs, goannas, owls, kookaburras, parrots, possums and tree-rats) nest or shelter in tree hollows.

Other animals may survive fire but suffer increased risks of predation in the more open burnt areas. Where fires occur repeatedly over many years, such species will decline. Conversely, single fires may also benefit some species. Many raptors (hawks and eagles) are attracted to burning areas to feast on fleeing prey or the casualties of fires. In unburnt areas, the grass layer may be so dense that many animals may have difficulty moving through it. Fires remove such obstacles, and allow readier access to resources (such as fallen seed) that have survived the fire. The occurrence and timing of fires may

① Saltwater crocodile, Shelburne Bay, Cape York Peninsula.
Photo by Kerry Trapnell

simple understorey structure, characterised by the dominance of annual grasses.

Each of these variants may favour particular suites of animal species. However, the current fire regime is such that these variants are far from balanced: there are remarkably few areas in which shrubby understories, and their associated fauna, are sustained.

Some animal species have complex requirements of fire. For example, Partridge Pigeons are most favoured when their territories include some areas that are burnt (where they can forage most efficiently) and some that are unburnt (where they prefer to nest in or under dense grass, and find some refuge from predation) (Fraser 2001). Within any year, the timing of fire will affect the timing of subsequent seed set in grasses. A seed-eating bird (or rodent) will be more advantaged when its home range includes some areas that are unburnt and other areas that are burnt at different times, for this situation will allow a far more extended period of seed availability than if its entire home range was either unburnt or burnt. Likewise, home ranges that include patches with different burning histories over a longer period (say 5–10 years) will also offer a broader menu of food and shelter resources than those exposed to a more homogenous fire regime (Woinarski *et al.* 2005). In most cases, animal species will benefit from fire patterns that are of similar scale to their home range. Current fire regimes are generally occurring at a far coarser scale.

INTERCONNECTIONS OF ECOLOGICAL MOVEMENTS

The northern landscape is crisscrossed with threads that connect any place with many other places. Water flows through landscapes; nutrients runoff and accumulate; plant seeds are blown or float from place to place; and fleshy fruits are moved by birds and bats.

Some movements, such as most of those listed above, are largely passive. Others are far more directional, forced or selected. For any animal species, the landscape offers a highly uneven and constantly shifting availability of resources. There are several ways of dealing with such inconstancy. Some species may shut down during the lean times. Some freshwater turtles spend much of the Dry season torpid, encased in

affect the productivity and timing of flowers and seeds of many plant species, and hence their availability for consumer animals. While single fires may have any of these impacts, fire regimes may have even more substantial impacts.

Taller shrubby understories develop in areas in which fire is excluded or infrequent. Many of these shrubs comprise species that produce fleshy fruits ('bush tucker'), which provide key resources for animals such as possums and the squirrel-like tree rats. In contrast, frequently burnt areas develop a more

and under the hard mud of dried up waterholes (Grigg *et al.* 1986; Kennett and Christian 1993).

Many frogs similarly aestivate under the ground for the Dry season. Many reptiles, including Frill-necked Lizards, remain inactive over much of the Dry season. Many plants shed their foliage and slow their growth rates over the course of the Dry season. Another strategy is to store food during the good times, and eat into that store during the bad, as squirrels store acorns. There are not many such cases in Northern Australia. Some ants harvest and store seeds during the late Wet and early Dry season (Andersen *et al.* 2000). Other species grow fat during the good seasons and this stored fat may tide them through periods of shortage. Rock-rats and dasyurids (small marsupial carnivores) may be the most conspicuous examples of this approach, developing greatly swollen tails at times of high food availability. Yet another approach to food shortage is to change diet to encompass whatever is locally available at any time. The most conspicuous example of this in Northern Australia is the marked increase in nectar-feeding employed by many otherwise typically insect-eating birds during the Dry season, with a special focus on the prolific and conspicuous flowering of the common dominant tree, Darwin Woollybutt, *Eucalyptus miniata* (Franklin 1999).

But by far the most common approach adopted by animals to the ebb and flow of resources is to move around the landscape, searching for habitats or patches where food is more available. This is by no means a strategy restricted to wildlife. Aboriginal people had a long history of undertaking seasonal movements across their clan estates, with these movements associated with shifts to areas where food was relatively more available and abundant (Russell-Smith *et al.* 1997). Pastoralists employ a similar strategy, with foraging cattle moving within and among paddocks to track food availability; and with some pastoralists shifting cattle between properties to overcome local shortage.

In the box on this page and the following page we describe some of the movement patterns that animals in Northern Australia use in order to cope with variation in food and other resources imposed by the strongly seasonal climate. These movement patterns range in scale from hundreds of metres to thousands of kilometres; in numbers from small family groups to hundreds of thousands of individuals; and in regularity from the clockwork to

WATER PYTHON & DUSKY RAT

The lower reaches of most of the large rivers of Northern Australia lie within extensive fertile plains. In the Dry season, these rivers are slow-moving, restrained within sinuous channels; and their surrounding blacksoil plains are cracked and baked dry. But during the torrential rains of the Wet season, the rivers break out of their channels and flood across vast areas of these low-lying flats. These floodplains are a simple but highly productive environment.

Two of the most characteristic species of these floodplain environments are the Dusky Rat *Rattus colletti* and Water Python *Liasis fuscus*. The Dusky Rat is a medium-large (to 220 g) terrestrial rodent, restricted to the Top End of the Northern Territory. The Water Python is a large (to 2.5 m) partly-aquatic snake. The two species are tightly linked in a simple food web.

There are two main features of interest in this system. The first is its dynamic. The rats don't, indeed can't, remain on the floodplains during inundation: they would drown. So, every year, they migrate (over a scale of hundreds of metres to several kilometres) from the floodplains up the shallow topographic gradient to 'upland' woodlands and forests. This is a relatively regular, predictable dispersal, without close parallel among any other Australian mammal. Keen on a diet of rats, the pythons make a comparable but more diffuse migration, for they can expand their Wet season diet to include more wetland resources, such as the eggs of waterfowl (Madsen and Shine 1996).

The other main feature of this system is its productivity. The density and biomass of rats in this system is extraordinary – rat numbers can exceed 600 individuals per hectare, and biomass approach five tonnes per square kilometre (Madsen *et al.* 2006). The density and biomass of pythons is similarly enormous.

This unique system depends upon the juxtaposition of floodplains and upland woodlands. The rats and pythons need access to both in order to survive. Such species may pose more substantial conservation challenges than species reliant upon only one habitat, for modification of either habitat may compromise their ecology.

John Woinarski

the chaotic. Some are entirely directional and predictable; others are more flexible (or desperate) attempts to respond to a capricious and unreliable patterning of resource availability.

As an example of biological movements, consider the geographic area across all of the North. Connect the threads of animal and plant movements here, and the pattern is remarkably

MAGPIE GEESE

The Magpie Goose *Anseranas semipalmata* is another feature species of the floodplains of Northern Australia, an icon more appreciated because it is far more conspicuous than the rats and pythons. It too may reach exceptional densities and biomass, especially during the late Dry season when the geese are concentrated on the relatively few and diminishing wetlands available.

The Magpie Goose is an important wildlife species in Northern Australia. It remains one of the major components of bush tucker in Aboriginal societies, and is the principal focus for recreational shooters. It is a taxonomic oddity, placed in a Family of its own, without close relatives. It is now mostly restricted to Northern Australia and southern New Guinea, though was formerly at least locally abundant in parts of south-eastern Australia. (The pattern of decline from southern Australia but maintenance of status in Northern Australia is noteworthy, and shared with many other species, such as Bush Stone-curlew, Red-tailed Black-cockatoo, and Australian Bustard.)

There are several features of the ecology of Magpie Geese that illustrate the significance of landscape linkages. Magpie Geese nest amongst emergent or floating vegetation of inundated floodplains. Their nesting requirements are fairly specialised: they need sufficient but not too much floodwater. A feature of the Wet season rains is that they show substantial spatial variability. Hence, in some years, the floodplains of the Daly River may be (most) suitable for Magpie Geese; whereas in other years the floodplains of the Mary River may be better. In some years, most floodplains may be entirely unsuitable. Magpie Geese respond to this broad-scale spatial variability by changing the spread of breeding colonies (and colony sizes) between floodplains between years. The maintenance of the Top End population of Magpie Geese is dependent upon the maintenance of options: the more floodplains there are to choose from, the more likely it is that, in any year, at least some floodplains will be suitable (Whitehead *et al.* 1992).

But Magpie Geese also depend upon finer-scale spatial linkage. The best place for Magpie Geese to position their nests is attached to robust emergent grasses and sedges, occurring in deep waters. But such sites may be drowned if the water levels rise too much; and they don't provide much suitable food for newly-hatched young, which must grow rapidly to escape the drying swamps before they become vulnerable to terrestrial predators. The best food sites for the young goslings are dense patches of seeding wild rice, and these typically grow in shallow waters often fringing the river channels. Soon after

Magpie Geese, Cape York Peninsula. *Photo by Kerry Trapnell*

hatching, the gosling brood must undertake a forced march (of up to 15 km per day) from nest to feeding ground. The juxtaposition and linkage of these different habitats within a single floodplain is critical for the breeding success for this species (Whitehead 1999; Whitehead and Dawson 2000).

Life for Magpie Geese is one long convoluted series of linkages. During the Dry season, the geese are dependent upon the bulbs ('corms') of a water chestnut *Eleocharis dulcis*. These subterranean corms are accessible to geese only when the water depth is about 30 cm or less, and cannot be accessed when the ground has dried hard. Over the course of the Dry season, swamps dry up rapidly. At any one swamp, this pivotal food resource is available only for a limited time: each swamp offers only a relatively narrow period for optimal feeding, and if all swamps developed and dried in synchrony, then the birds would be offered a brief glut, followed by a prolonged famine. To survive the Dry season, these birds need to move between a network of swamps across a broad landscape. The intricate natural balance on which this species depends is increasingly destabilised by the spread of exotic pasture grasses to its wetland habitat, with these capable of displacing the native plants that provide key food resources at critical times in its life history (Whitehead and Dawson 2000).

John Woinarski

complex: this piece of land (or any other) has biological connections to many other local regions, to distant regions of the continent, and to other continents. For part of every year the shorelines and waterbodies in this area will be home to shorebirds (waders) that spend the rest of the year in northern Asia. For part of the year, the rainforests and savannas here will be home to cuckoos and koels, dollarbirds, swifts and bee-eaters, which spend the rest of the year in New Guinea, Indonesia or Malaysia. For part of the year, the savannas will be home to birds such as Black Kites and Grey Fantails that spend the rest of the year in central or temperate Australia. For part of the year, whales will live in these waters and then disperse to colder seas; barramundi, sawfish and crocodiles will move between the fresh waters and the seas. Flying-foxes and rainforest pigeons will move among the network of isolated rainforest patches, depending on which patch happens to provide

the most fruit at any time (Figure 3.4). Driven by the locations of rainfall and waters, Magpie Geese will move throughout this landscape, with different critical sites at different times of year. Black Cockatoos, kites and other birds will move across this entire landscape in different patterns every year, as they track burnt areas. Other species, such as quails, may try to track unburnt areas. At the onset of the rains, termites and flying ants will disperse to found new colonies; butterflies will disperse from Dry season refuges to expand their distributions in the Wet season. The movements are regular, chaotic, predictable, indecipherable, solitary, in massed aggregations, local, regional, international.

All movement patterns are significant. In some cases, the dispersal of animals provides for the dispersal of the plants on which they feed (such as by fruit-eating flying-foxes and pigeons). All demonstrate that the ecological fabric of

FIGURE 3.4 **AN EXAMPLE OF LONG DISTANCE MOVEMENTS BY WILDLIFE**

The roosting locations of a single Black Flying-fox over a three-month period near Darwin. Black Flying-foxes pollinate flowers and disperse seeds of many savanna and rainforest trees.

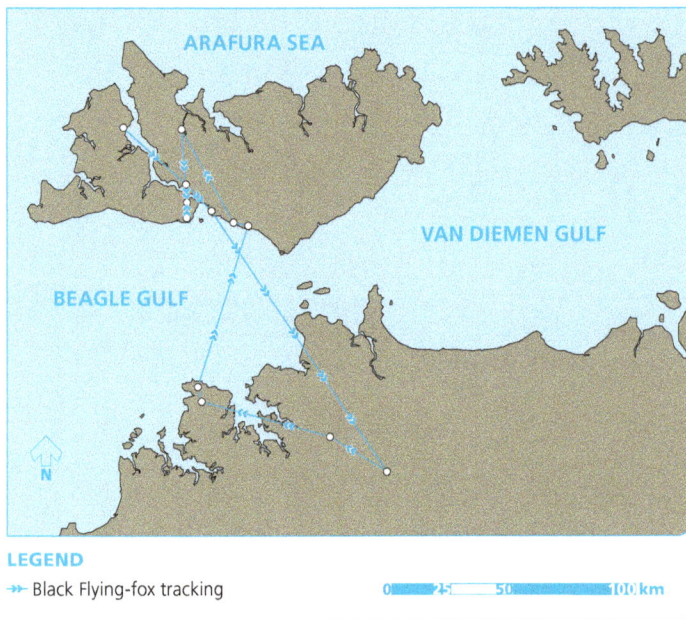

LEGEND
→→ Black Flying-fox tracking

0 25 50 100 km

Source: Carol Palmer, NT Natural Resources, Environment and the Arts (NRETA).

Northern Australia is woven from myriad interconnections. Removal or degradation of any site will have repercussions elsewhere in the landscape; the viability of any place is dependent upon the fate of many other places. Most conservation reserves will not alone protect these mobile species, for the scale of their movements is often larger than that of single reserves, and the set of reserves will rarely be positioned to match their regional movement patterns.

OTHER CRITICAL ECOLOGICAL PROCESSES AND LINKAGES

In the section above, we provided examples of the workings and importance of three main ecological processes in Northern Australia. There are at least four other processes that are important, but are not considered here in such detail.

Strongly interactive species

Some individual species have a disproportionately major impact on the community of species in which they live, with this influence working across a range of scales. Such species include: (i) major predators (such as dingoes and green tree ants) which may control the relative abundance of prey species (and hence the structure of plant communities) on which they depend; (ii) animal species (such as flying-foxes, pigeons and fruit-doves) critical for the dispersal of the fruits or seeds of plants; (iii) species that change the dynamics or structure of habitats (such as some termites, which are vital for the formation of the tree hollows used by many other animals); and (iv) 'keystone' species that provide resources for many species, particularly at times when few other resources are available (such as some figs, and cockatoo grass). Decline or loss of such species is likely to have impacts that percolate widely across the landscape.

Climate change and variability

The world is an inconstant place, and this inconstancy may drive evolutionary divergence, and the prosperity or extinction of different species. The climate of Northern Australia, along with other aspects of its geography has changed over time; for example, with the inundation of the former land bridge connecting New Guinea, Cape York Peninsula and Arnhem Land, some 6000–8000 years ago (Mackey *et al.* 2001). Accelerating rates of climate change are likely to ratchet up the ecological and evolutionary pressures on species in Northern Australia (as elsewhere), compromising the viability of some. Temperatures and sea levels will rise, rainfall patterning will change, and there may be an increase in extreme weather events (notably severe cyclones) and in fire severity. Most at risk are those species currently coupled to a narrow climate preference or a fixed juxtaposition of habitats. The conservation challenge is to identify such species, and maintain as many landscape options as possible for them. In general, the more the landscape is fragmented and developed, the fewer the options for such species.

Land-sea connections

Northern Australia has a long coastline and marine influences reach far upstream in the large tidal rivers. In these blurred boundaries, mangrove forests are extensive, diverse and productive. They have significant linkages with marine and terrestrial systems, perhaps most notably as key breeding areas for many important fish species. Mangrove habitats also provide the primary habitat for a range of – typically highly specialised – plants,

① Dingoes may control populations of large kangaroos and feral cats in the North. *Photo by Lochman Transparencies*

invertebrates, reptiles (mangrove snakes and mangrove monitors) and birds.

Shorebirds, seabirds, many fish, crocodiles, mangrove snakes, marine turtles and many other species depend upon the coastal frontier and/or need to move between land and sea.

Much marine productivity and functioning depends on inputs from the land, and the health of those terrestrial systems may determine the health of adjacent and more distant marine systems. For example, the productivity of the northern prawn fishery is closely related to rainfall patterning and the consequential amount of nutrients transported in rivers.

Northern Australia has thousands of islands, including (after Tasmania) Australia's second (Melville Island), third (Groote Eylandt) and fifth (Bathurst Island) largest islands. Isolation has sheltered these islands from many destructive processes that have detrimentally affected mainland species and environments. Many now contain plant or animal species no longer found on the mainland (Southgate *et al.* 1996; Woinarski *et al.* 2000a, 2003). Many provide glimpses of what the Northern Australian mainland may have looked like before the influence of European colonisation.

Maintaining evolutionary processes

Connections and refuges across landscapes allow for long-term changes in the range of species, the genetic flow within species across this range, and the evolution of new species. For example, during the colder drier times of the repeated Ice Ages, refuge areas of tropical rainforest survived on Cape York Peninsula. The rainforest expanded out of these areas during warmer, wetter times, and may contract again in the future. Destruction or fragmentation of refuge areas could prevent such processes happening in the future.

CONCLUSION

In this chapter we have considered the major natural processes that make the landscapes of Northern Australia work. These processes, especially disturbance (and especially fire), hydro-ecology and long-distance movement of species through the landscape ensure that ecosystems are sustained.

Building upon this knowledge, in Chapter 4 we examine the natural values that flow from and are dependent upon the ongoing healthy functioning of the North's landscapes.

①

THE NATURAL VALUES OF NORTHERN AUSTRALIA

For many Australians, the overwhelming impression of Northern Australia is of nature in abundance: endless tracts of savanna, flocks of Magpie Geese sweeping down to a billabong, vast wetlands swollen from heavy rains. On a planet where nature is often the pieces left behind after intensive human modification, the North is a place where nature stands out.

In the previous chapter we discussed the ecological processes and connections that maintain these natural values of the region. Here we give an overview of these values and their significance, globally and nationally. We do not attempt to describe the full suite of values that occur in all the regions across the North.

We focus in this chapter on describing some specific assessments of the national and international significance of the natural values of Northern Australia. These frames of reference are important because it is easy to take the local and familiar for granted; what is locally commonplace may be nationally or internationally of great significance.

LOCAL VALUES

There are discrete localities in Northern Australia that are exceptional by any measure. Their value is clear and generally well-understood and acknowledged. Two of the more obvious such examples are described in some detail in the boxes *'The Arnhem Land Plateau'* on page 46 and *'Life along land's edge – the wading birds of Roebuck Bay, Broome'* on page 47. Other obvious cases of outstanding values for biodiversity conservation include the North Kimberley and its islands (particularly so for endemic plants and animals, and for mammal species that have declined across most of the rest of their range: Wheeler *et al.* 1992; Graham and McKenzie 2004) and Cape York Peninsula (Abrahams *et al.* 1995; Mackey *et al.* 2001).

Change the scale and focus marginally and there is a vast network of sites across Northern Australia that have national significance for biodiversity, because they maintain populations of threatened species, large assemblages of waterfowl or other wildlife aggregations, important nesting sites for seabirds or marine turtles, unusual species richness, or support

1 Nesbit River, McIlwraith Range, Cape York Peninsula. *Photo by Kerry Trapnell*

THE ARNHEM LAND PLATEAU

Loosely, we can characterise Northern Australia as comprising three main environmental elements: the vast expanse of low-land plains, typically supporting eucalypt woodlands; a more discrete network of highly productive wetlands; and the stone country, a series of rugged rocky outcrops and ranges that give the land its grandeur and sense of immemorial time. Each has a contrasting biota; each its own problems.

The rugged uplands are most extensive and variable in the Kimberley, but there are other significant ranges scattered across the breadth of Northern Australia.

For cultural and natural values, the Arnhem Land plateau is one of the most important of these ranges. This predominantly sandstone massif occupies about 32,000 km², of which about one quarter is included within Kakadu National Park. Although it contains spectacular cliffs, escarpments and gorges, the total altitudinal range is relatively limited: the highest point is only about 400 m.

The caves and sandstone walls of this plateau contain some of the most spectacular and abundant rock art in the world, and this stone country is densely populated with sites of major cultural significance to Indigenous people.

The Arnhem Land plateau is extremely rich in endemic (restricted) plants and animals. About 200 plant species occur nowhere else in the world (Woinarski *et al.* 2006a). Endemic animals include three birds (the White-throated Grass-wren, Chestnut-quilled Rock-pigeon and Banded Fruit-dove), 12 reptiles (including one of Australia's largest snakes, the Oenpelli Python); five mammals (including the rare Black Wallaroo); together with three fish, one frog and numerous invertebrates. These latter include a spectacular radiation of crustaceans, including an endemic family of shrimps (Kakaducarididae) and exceptional species diversity in the *Eophreatoicus*, an endemic genus of isopods. Many of these species are extraordinarily localised, to single streams or springs (Finlayson *et al.* 2006).

The high levels of endemism and species richness arise from a combination of factors. For some species-groups (such as the trigger-plants *Stylidium* and resurrection grasses *Micraira*) with limited dispersal ability, the deep gorges, spectacular waterfalls and/or sheer cliffs isolate local populations, ultimately giving rise to divergence and speciation. The topographic variability of the stone country also promotes richness, through offering such a variety of microclimates and microhabitats. In contrast to the surrounding lowlands, the plateau has stood largely

Arnhem Land escarpment. *Photo by Glenn Walker*

stable in the landscape for more than 100 million years, and has provided refuge from (or has moderated the impacts of) climate change, inundation and fire. Many species have persisted here since ancient times, and are now relictual markers of an earlier age: such species include the plants *Drummondita, Hildergardia* and *Podocarpus*). The most notable of these relicts is the large evergreen tree *Allosyncarpia ternate*, a primitive relative of eucalypts, that now dominates the patchy network of rainforests scattered across the gorges and cliffs of the Arnhem Land plateau.

The ruggedness of the plateau has provided protection to these plants and animals since ancient times. Over the last 50 years, that protection is breaking down. The plateau lands are now largely depopulated, and the loss of traditional fire management has resulted in what is now an anarchic regime characterised by frequent extensive fires. In the absence of managers, feral animals (especially Water Buffalo) have spread to all but the most inaccessible parts of the stone country. The distinctive and finely tuned ecological communities of the stone country are unravelling. What was for so long inviolable and sheltered is now exposed: the stone country now has more threatened species than anywhere else in the North.

John Woinarski and Brendan Mackey

stable or increasing populations that are in decline across most of the rest of their range.

To a large extent, there has been a detailed accounting of such sites over most areas of Northern Australia over the last few decades, notably so for the Kimberley (Burbidge *et al.* 1991), Northern Territory (Anon. 2007) and Queensland (Stanton 1976; Abrahams *et al.* 1995; Mackey *et al.* 2001). We do not aim to repeat such analyses here.

Instead, our approach recognises that such sites are indivisibly connected to the landscape as a whole. The maintenance of their significant site-specific values is dependent upon the continuing connectivity of landscape-wide ecological processes. For example, the status of conservation assets on the Arnhem Land plateau is affected by fires coming from the surrounding lowlands. The numbers of wading birds visiting Broome will decline if wetlands and coastal areas elsewhere become degraded.

Further, as described in the opening paragraphs of this chapter, we aim here to see this land in a broader national and international context: what is the conservation significance of Northern Australia as a whole? Thus, we choose not to divide and compartmentalise this landscape into artificial segments and weigh up the measure of each of these individually – the region's value is far more than the sum of its constituent parts.

NATIONAL AND INTERNATIONAL VALUES

The state of the world's tropical savannas

Tropical savanna occurs in tropical monsoonal climates throughout the world, and once covered about 12% of terrestrial Earth (16.1 million km²). The most extensive areas occur, or occurred, in Northern Australia, Africa, India and South America. Smaller areas are found in Madagascar, Indochina, Indonesia, the Philippines and southern New Guinea (Figure 4.1). The different savannas vary greatly in their composition of animal and plant species present but, due to the similar climates of alternate Wet and Dry seasons, have similar vegetation structure – sweeping grasslands with varying levels of tree cover, sometimes with a generally open shrub layer.

LIFE ALONG LAND'S EDGE – THE WADING BIRDS OF ROEBUCK BAY, BROOME

Every year millions of shorebirds – sandpipers and knots, godwits and curlews, whimbrels and tattlers – come to Australia from the Northern Hemisphere. Their life is spent chasing an eternal summer.

Eastern Curlew flock
Photo by Dean Ingwersen

In our winter they breed in the tundra, taiga and shores of Asia and North America. During our spring they head south, seeking the southern summer on the shores of Australia. After summering here, they moult into breeding plumage, store fat and head North again in our autumn, off to their breeding grounds, up to 10,000 km away. This great bi-annual river of birds is called the East Asian-Australasian Flyway.

Throughout the world, shorebirds rely heavily on very specific habitats. Most species prefer marine mudflats in their non-breeding seasons. These are rich in shellfish, crabs and marine worms. Northern Australia has many areas of such habitat with important wader habitat in areas such as Darwin Harbour, and the southern coast of the Gulf of Carpentaria.

However all mudflats are not equal. A shallow sloping coast, soft muddy sediments and big tides are necessary to produce huge mudflats that make prime habitat for hundreds of thousands of shorebirds. There are only a dozen or so areas in the world with such huge intertidal mudflats rich in shorebirds.

The Kimberley has two of these global shorebird hotspots – Roebuck Bay, and just to the south, the 80 Mile Beach. These are by far the richest shorebird habitats in Australasia. And only one other such shorebird-rich area occurs in the tropics – in the Guyanas in South America.

Every year nearly a million shorebirds of more than 25 species migrate to Roebuck Bay and 80 Mile Beach. They join a few species of resident Australian shorebirds that never leave the continent.

Northern Australia also has the third most important shorebird area in Australia – the mudflats and mangroves of the southern coast of the Gulf of Carpentaria.

Concentrations of shorebirds can be easily seen by visitors on Roebuck Bay at the Broome Bird Observatory near Broome town, and also on the Esplanade in the middle of Cairns on the other side of Northern Australia.

Barry Traill

COMPARING THE STATE OF THE WORLD'S TROPICAL SAVANNAS

To assess the condition of tropical savanna woodlands globally, we combined data on three factors associated with effects on the vegetation cover of a landscape:

- The proportion of land cleared for cropping (Figure 4.1a) – this represents an extreme end of the gradient of native vegetation condition, indicating the complete or near complete removal of natural vegetation from a landscape;
- The density of livestock, such as cattle, sheep and goats (Figure 4.1b) – the effect of this factor depends on the carrying capacity of the environment but, in the absence of comprehensive data on carrying capacities, we assumed that the higher the density of livestock, the more likely a landscape was to be exposed to unsustainable grazing regimes; and
- The density of the human population (Figure 4.1c) – potentially degrading activities (such as over-cutting of fuel wood for heating and cooking) are associated with large numbers of people, particularly among subsistence farmers and pastoralists.

Further details of the data sources and methods used in this analysis are given in Appendix 1.

Combining the data from the above factors gave a vegetation disturbance index (Figure 4.2). The values of this index range from 0.0–1.0. Low values indicate poor condition due to extensive cropping, high densities of livestock, and large human population numbers; high index values represent the opposite.

In interpreting this analysis, we recognise that cropping, livestock density and human populations are not the only factors affecting savanna health. For example, tropical Australia contains large numbers of feral herbivores, including water buffalo, pigs and cattle. These animals can degrade the vegetation cover in particular parts of the landscape, such as around water holes, during the Dry season (see Chapter 5 for further discussion). Although Australian savanna woodland has a relatively low density of livestock, the total density is not the sole determinant of environmental impact. Where livestock are concentrated in the landscape they can lead to soil erosion; for example by destroying natural levee banks, which leads to rapid draining of areas, drying of soil and loss of vegetation (Pringle and Tinley 2003). Also, as explained above, effects of livestock density depend on carrying capacity of the landscape.

This index also assumes that the greater the human population in an area, then the more degraded the vegetation condition. However, in many parts of Northern Australia there are probably fewer people living on country than at any time in the last 50,000 years (the earliest recorded data for human occupation of the continent). Some of the less obvious ecological problems emerging in Northern Australia are related to Aboriginal depopulation over the last century and some of the solutions to the problems are to be found in re-establishing active management of the land.

① Eucalypt savanna, Port Stewart, Cape York Peninsula. *Photo by Kerry Trapnell*

A recent global analysis *(The United Nations Millennium Ecosystem Assessment: Millennium Analysis)* examined the natural integrity of the world's biomes (large areas of similar climate and vegetation).[1] Although that study did not specifically consider savanna or savanna woodland, it assessed the state of 'Tropical Dry Forest' and 'tropical and subtropical grasslands, savanna and shrub lands' ('Tropical Grasslands'). The Millennium Analysis showed that Tropical Dry Forests have been significantly affected by cultivation, with almost half of the biome's native habitats replaced by cultivated lands. Tropical Dry Forests and Tropical Grasslands have both had around 60% of their natural distribution cleared by humans, with about 15% of this occurring since 1950. The Millennium Analysis also noted that in addition to the total amount of habitat loss, the spatial configuration of loss can strongly affect biodiversity, with habitat fragmentation typically accompanying land use change, leaving a complex landscape mosaic of native and human-dominated habitat types. The Millennium Analysis highlighted that habitat fragmentation typically endangers species by isolating populations in small patches of remaining habitat, rendering them more susceptible to genetic and demographic risks as well as natural disasters.

Building on the Millennium Analysis and other studies, we analysed the state of the world's tropical savannas, as context for the savannas of Northern Australia. We used computer-based geographical information from a variety of international sources, including NASA satellite sensors (see box *'Comparing the state of the world's tropical savannas'* on page 48 for details of this study). The global datasets we used covered tropical savanna woodland, defined here as 'savannas with trees'. The analysis therefore did not cover treeless tropical savanna grasslands, shrublands, and denser tropical forest.

The results are presented in Figure 4.1, which provides a combined index of the extent to which these savannas vary from relatively natural to highly modified by human activity.

Globally, tropical savanna woodland originally covered 11.99 million km². Almost 70% of these savanna woodlands have now been removed, leaving only 3.69 million km². Of the remaining savanna woodlands, only about 22% has a high vegetation condition index (shown in blue in Figure 4.2 and Figure 4.3), and

FIGURE 4.1a **WORLD SAVANNA CONDITION, CROPS**

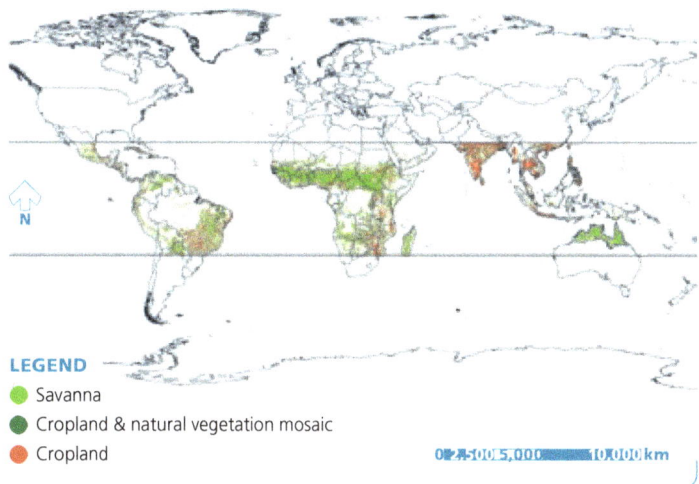

LEGEND
- Savanna
- Cropland & natural vegetation mosaic
- Cropland

FIGURE 4.1b **WORLD SAVANNA CONDITION, LIVESTOCK**

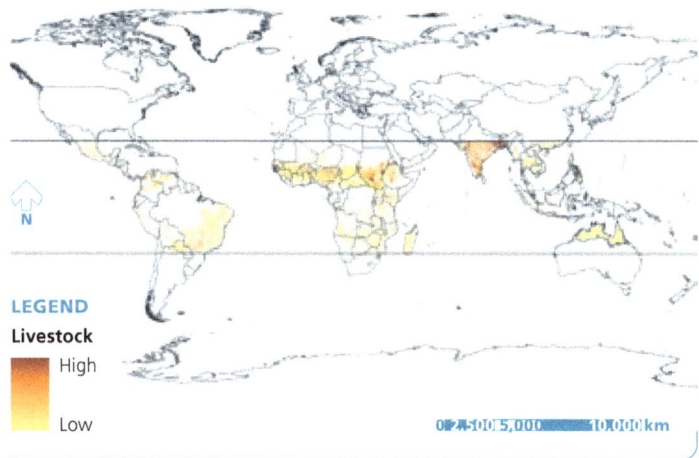

LEGEND
Livestock
- High
- Low

FIGURE 4.1c **WORLD SAVANNA CONDITION, POPULATION**

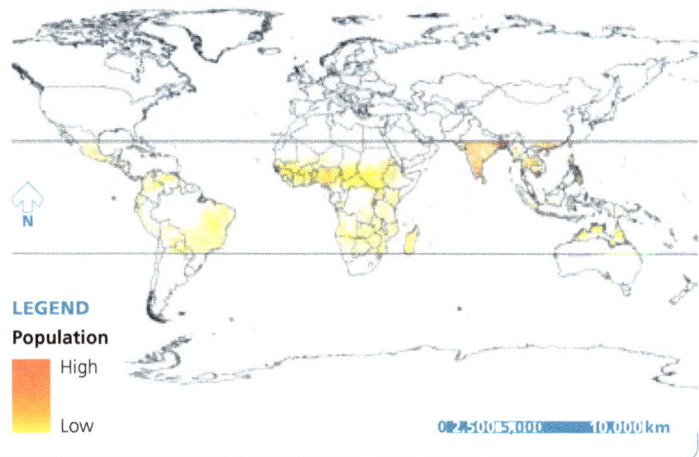

LEGEND
Population
- High
- Low

---

1 www.maweb.org/documents/document.354.aspx.pdf

FIGURE 4.2 **OVERALL WORLD SAVANNA CONDITION**

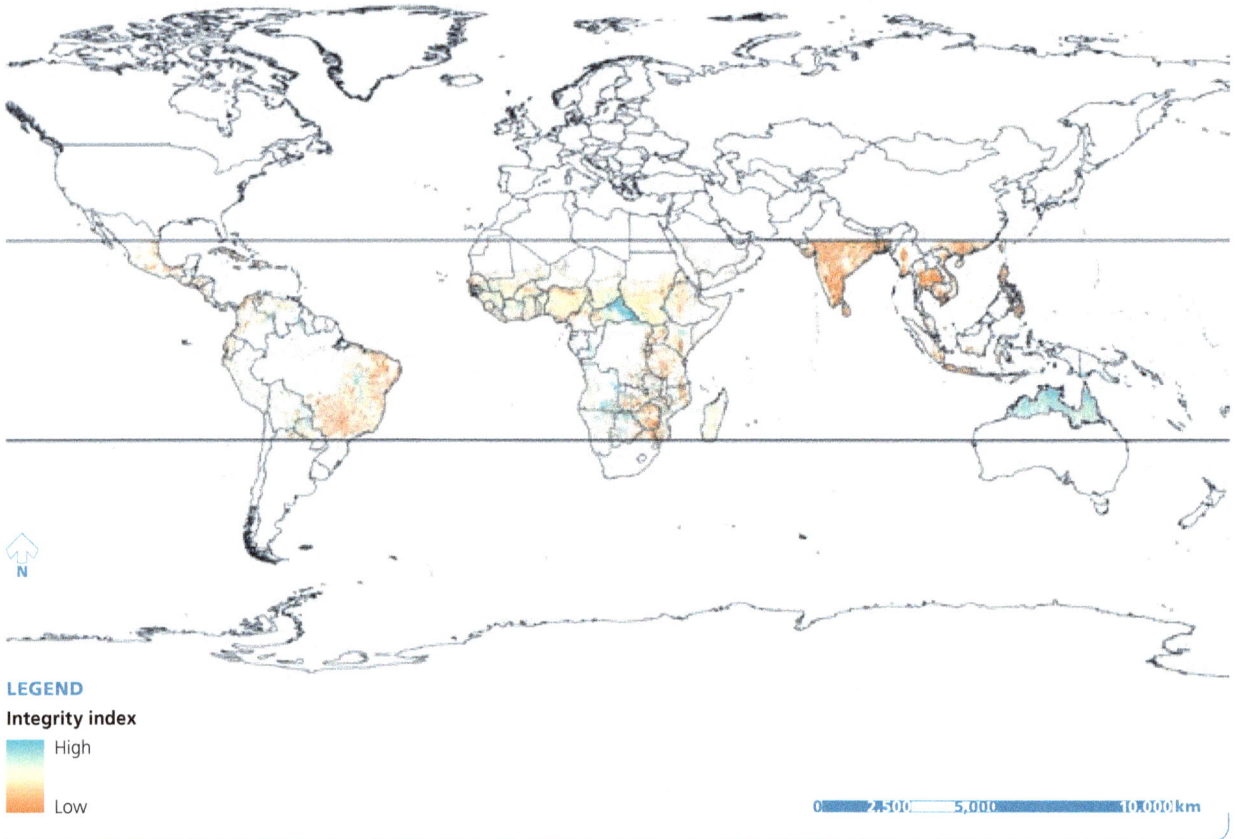

LEGEND

Integrity index

High

Low

N

0 2,500 5,000 10,000 km

① Jabiru on Gulf Country wetlands. *Photo by Wayne Lawler*

② Savanna grassland, Cape York Peninsula. *Photo by Kerry Trapnell*

①

much of this, especially in Africa, comprises fragmented patches rather than extensive areas.

By far the largest expanses of tropical savanna woodland remaining in good condition are in Northern Australia. Australia has more than 25% of the remaining savannas; no other country has more than 9%. Australia is also the only economically developed and politically stable nation containing extensive areas of tropical savanna. This combination of attributes makes Australian tropical savannas of very high conservation value at a global level.

This level of international importance of Australia's tropical savannas is in contrast to that of Australia's celebrated tropical rainforests, for which Australia contributes only about 1% of the world's tally.[2]

2 As of 1999, the extant area of tropical forest globally was estimated to be around 1,407,649 x 1.03 km², while in Australia, it was estimated at 14,088 x 1.03 km² (see EarthTrends online database of the World Resources Institute; www.wri.org).

FIGURE 4.3 **AUSTRALIAN SAVANNA CONDITION**

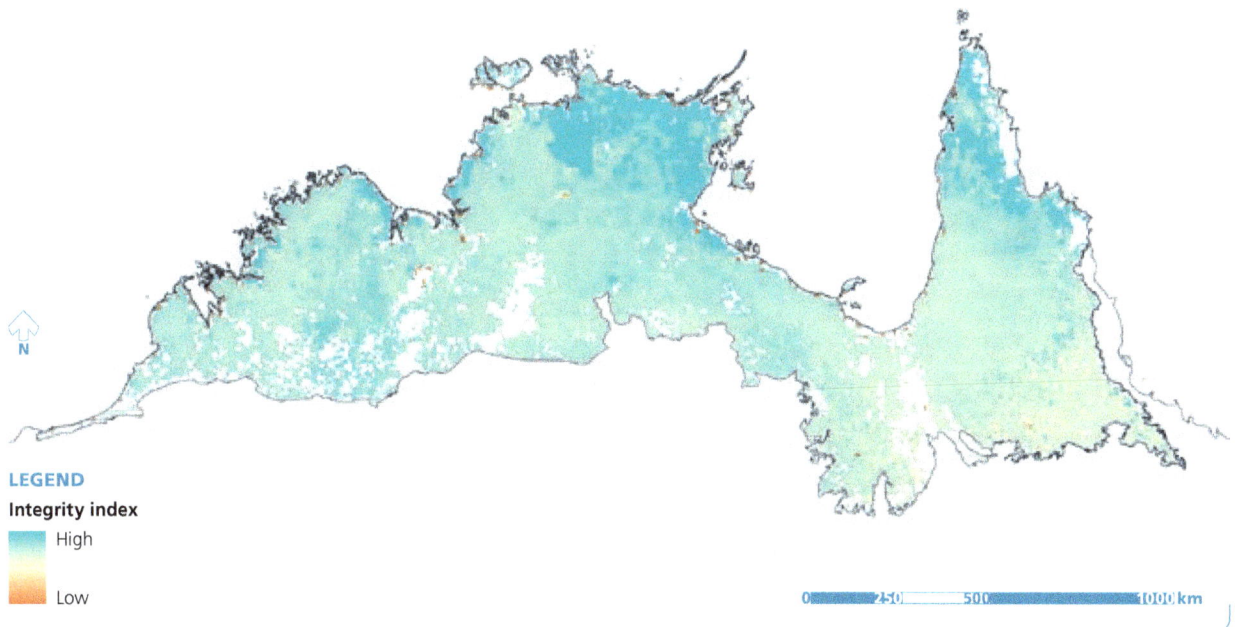

N

LEGEND

Integrity index

High

Low

0   250   500   1000 km

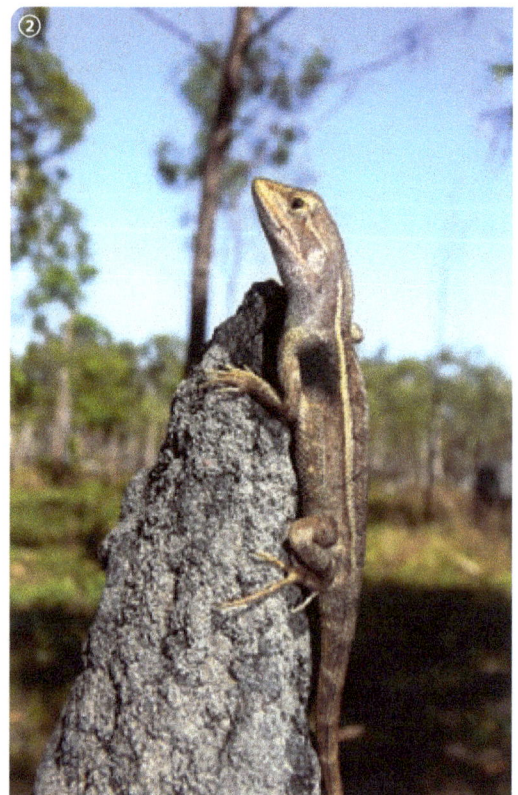

COMPARING THE STATE OF AUSTRALIA'S WOODLANDS

Vegetation structure (i.e. the height, density and layering of the plants) varies along a gradient from dense forest to grassland. The structure of tropical savanna woodland, as defined for our global analysis, mainly correlates with vegetation types mapped in Australia as 'woodland' (Figure 4.4). Given this, we compared the pre-1788 and current continental distribution and condition of Australia's woodlands, including the tropical savanna of Northern Australia, and the subtropical and temperate woodlands of Queensland, New South Wales, Victoria, South Australia and Western Australia.

Data on the extent of woodland and forest cleared for cropping is from 'Integrated Vegetation Cover V1' (BRS 2003). The 'replaced' class indicates woodland and forest cleared for a land use other than cropping. 'Modified' represent an intermediate level of change, which in the pastoral zone largely reflects the intensity of commercial grazing.

Details of the data and methods used in this analysis are given in the Appendix.

FIGURE 4.4 **DISTRIBUTION & CONDITION OF EUCALYPT WOODLAND**

To improve map interpretation, the relatively small areas of forest are shown as areas over-shadowed with dark grey. All other coloured areas are, or were, eucalypt woodland. White areas are other vegetation types. 'Residual' refers to woodland and forest that is the least disturbed from contemporary post-European settlement land use activity.

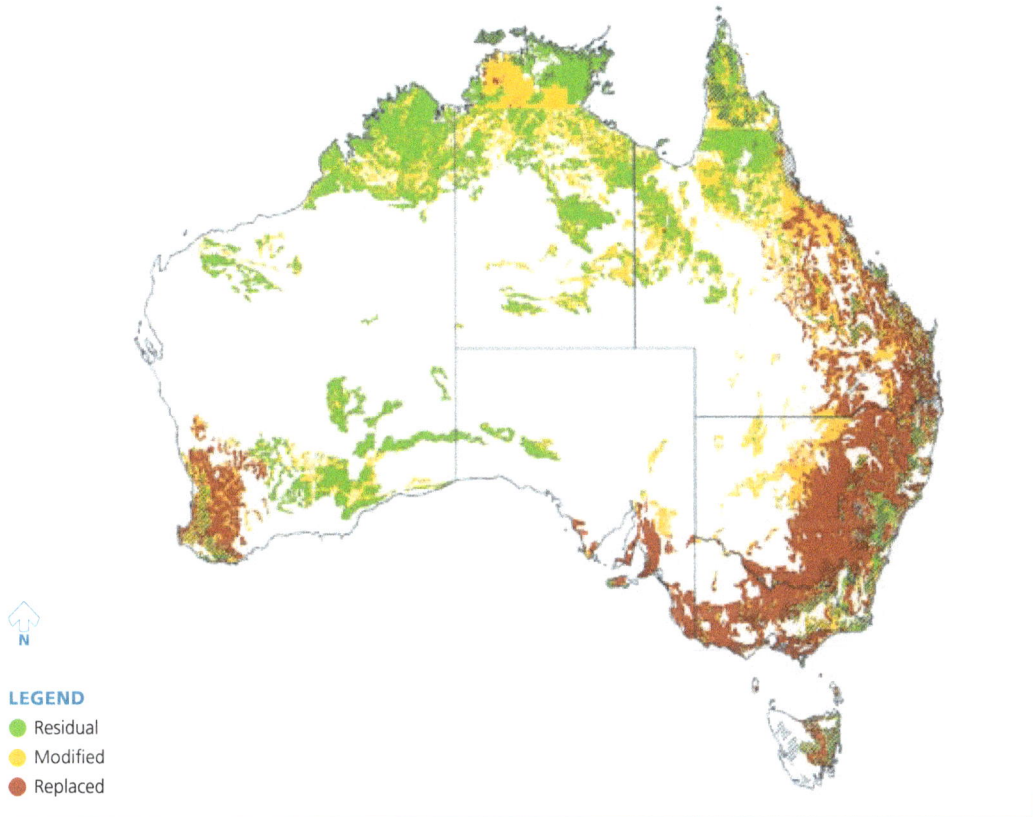

N

LEGEND
- Residual
- Modified
- Replaced

THE STATE OF AUSTRALIA'S WOODLANDS

We undertook a comparable analysis at a national scale of the significance of the dominant vegetation of the North–the savanna. 'Woodland' and 'forest' rather than 'savanna' is the broad term usually used in identifying and mapping all open-treed vegetation throughout Australia. Here we compare the condition of eucalypt woodlands throughout Australia. The procedures and data sets used are presented in the box *'Comparing the state of Australia's woodlands'* on page 52.

Figure 4.4 shows the distribution of Australian eucalypt woodland and forests before European settlement and now. Note that unmapped parts of Northern Australia (white space) in Figure 4.4 comprise other vegetation types, including savanna grassland and heathland.

More than 80% of the temperate and subtropical woodlands have been cleared for intensive land uses, or heavily modified by intensive grazing or other disturbances. In striking contrast, woodlands and forests remain extensive and in relatively unmodified condition in Northern Australia.

THE STATE OF AUSTRALIA'S WATERWAYS

Hydro-ecological processes are critical to the healthy functioning of country in the North (Chapter 3). Here we briefly examine the state of wetlands and waterways in Northern Australia relative to that of the continent as a whole, based on recent national and regional level analyses.

① Land clearing has removed more than 85% of Eucalypt woodlands in Southern Australia. *Photo by Barry Traill*

② A Diporiphora dragon, one of many species dependant on savanna. *Photo by Ian Morris*

FIGURE 4.5 **LOCATIONS OF LARGE DAMS IN AUSTRALIA**

Large dams – with a crest height greater than ten metres – marked with dots or numbers for shaded areas indicate numbers of dams, blue lines identify river basins.

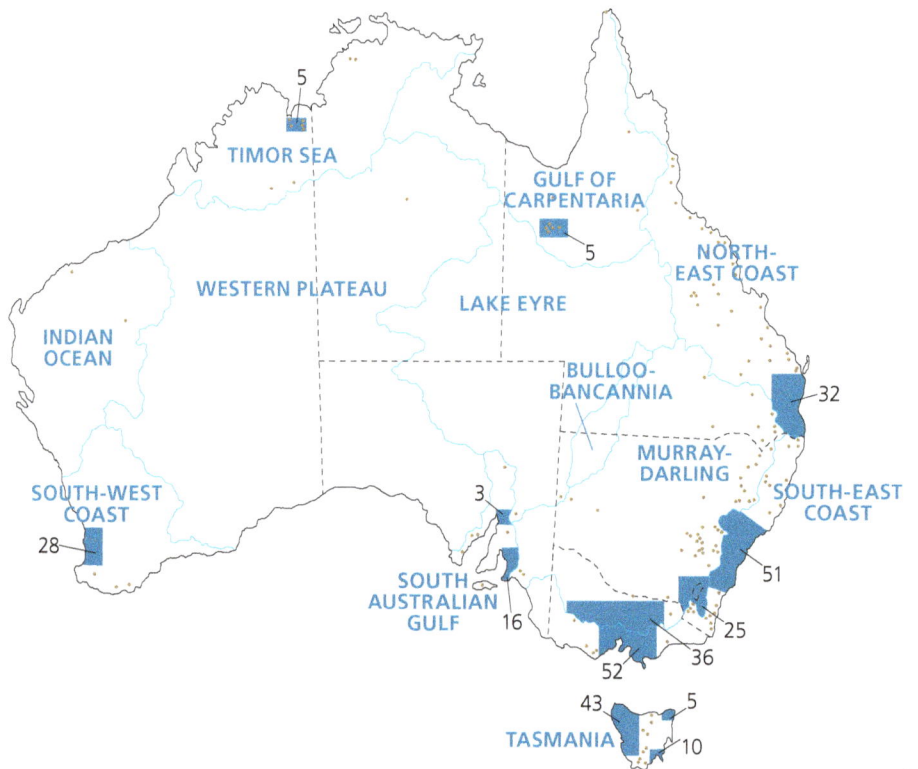

Source: Kingsford (2000).

Impacts of dams

One of the most significant changes humans can impose on the hydro-ecology of a landscape is the construction of large dams on major rivers. Australia has at least 446 large dams (crest height >10 m), which divert water from floodplain wetlands (Figure 4.5).

Dams can change aquatic vegetation, reduce vegetation health and reduce populations of waterbirds, native fish and invertebrates (Kingsford 2000). They can also cause some floodplain wetlands to become permanent storages, so that the naturally occurring plants and animals are replaced with species not tolerant of variable flooding regimes. Table 4.1 shows the number of large dams in water basins, and the percentage of divertible flow taken from each. Northern Australia has relatively few dams compared with other climatically humid water basins in Australia. Consequently, more of its floodplains and associated wetlands are intact.

An example of the potential impacts of large dams can be observed in the one of Australia's biggest dams, which is in the North: the Ord River Irrigation Area (ORIA). ORIA is a 150,000 hectare 'flow-through' irrigation scheme on the Ivanhoe and Packsaddle Plains near Kununurra. Irrigation waters diverted from Lake Kununurra are delivered by a gravity-feed system of channels. Agricultural effluent returns to the lower Ord River (via a drainage network). Maintenance of the 'hydraulic head' of water pressure, which is necessary for irrigation, restricts downstream water-flow patterns. Environmental impacts in the ORIA include high sediment loads due to sheet and gully erosion in the Ord River catchment, which reduce the water storage capacity of Lake Argyle; and inappropriate land and water management practices, which have effects on health and resource management. Water impoundment has reduced the distribution and abundance of barramundi to about one-quarter of their former range in the river (Doup and Pettit 2002).

Condition of rivers and catchments

The condition of Australia's rivers and catchments, including those of Northern Australia, has been comprehensively assessed in a recent study (Stein *et al.* 2002). This study calculated a 'River Disturbance Index', a measure of the impact of structures and human activities on river condition, in addition to

TABLE 4.1 NUMBERS OF LARGE DAMS & PERCENTAGE OF DIVERTIBLE FLOW

River basin	Number of dams (crest height >10 m)	Percentage of divertible flow (taken for human use)
North-east coast	63	16
South-east coast	128	28
Tasmania	70	9
Murray-Darling	107	81
South Australian Gulf	25	44
South-west coast	31	13
Indian Ocean	2	9
Timor Sea	9	9
Gulf of Carpentaria	8	1
Lake Eyre	2	13
Bulloo-Bancannia	0	0
Western Plateau	1	0
Total	446	21

Source: Kingsford (2000)

the impacts caused by large dams. As with the vegetation condition index, a high value represents a high level of disturbance.

Figure 4.6a shows the River Disturbance Index for Australia. This map may be misleading, because many of the rivers are in arid Australia and do not support regular water flows. The exception is the Lake Eyre Basin, where water that originates in the summer monsoon rains of Northern Australia causes seasonal flooding in the channel country, occasionally reaching Lake Eyre. For this report, we compared the rivers of Northern Australia with other climatically humid zones, where a surplus of water is generated to support stream flow and groundwater recharge. Figure 4.6b shows these rivers in the humid zones and their disturbance index values.

FIGURE 4.6a **RIVER DISTURBANCE INDEX FOR AUSTRALIA**

Class 1 indicates rivers and catchments with a relatively high level of natural integrity. Class 10 indicates rivers and their catchments that have been highly altered and disturbed by human activity, including water impoundment and diversion, and degradation of the catchment's vegetation cover.

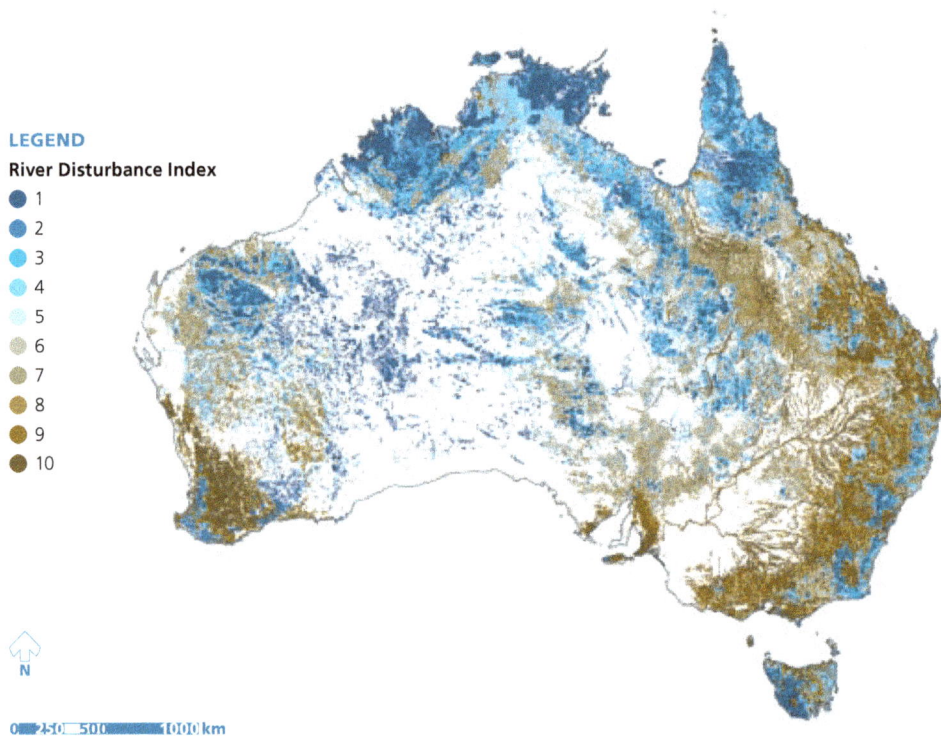

LEGEND

River Disturbance Index

- 1
- 2
- 3
- 4
- 5
- 6
- 7
- 8
- 9
- 10

N

0 250 500 1000 km

Source: Updated version of continental analysis by Janet Stein and colleagues (Stein *et al.* 2002).

From a continental perspective, it is clear that Northern Australia retains the largest expanses of intact rivers and catchments in the continent. Across Northern Australia, the land cover of most water catchments remains in relatively good condition and there has been little alteration to the flow regime from impoundments, flow diversions or discharges and levee banks. Natural river processes associated with hydrological, geomorphological and biological activity remain largely intact across most of Northern Australia, with profound implications for the conservation of biodiversity and associated natural values.

For Southern Australia, the general picture is of major degradation to river systems, with most rivers in poor condition. There remain some rivers with a high level of natural integrity, but these are largely restricted to some forested landscapes in mainland southern Australia and south-west Tasmania.

Any detailed assessment of the characteristics of the rivers of Northern Australia is hampered by the lack of suitable data. The Australian government undertook a data audit of Australia's northern rivers, including associated wetlands, estuaries and floodplains within a catchment and land use context (NGIS 2004). One main conclusion was that information about water resources is limited and incomplete for large parts of remote Northern Australia. For example, data are insufficient to understand disturbance and point-source pollution in the region, existing riparian (river bank) vegetation data are at too coarse a scale, and spatial data about inland fish is lacking. The lack of fine-resolution data means that certain environmental impacts on river condition (including the impact of feral animals) could not be factored into the River Disturbance Index.

The general picture of good condition of rivers in Northern Australia relative to those elsewhere on the continent is also supported by studies that

FIGURE 4.6b **RIVER DISTURBANCE INDEX WITHIN THE HUMID CLIMATIC ZONE**

Index classes as for Figure 4.6a. Shown only are river disturbance conditions for humid areas where a surplus of water is generated to support stream flows and groundwater recharge. The boundary of Northern Australia is also shown.

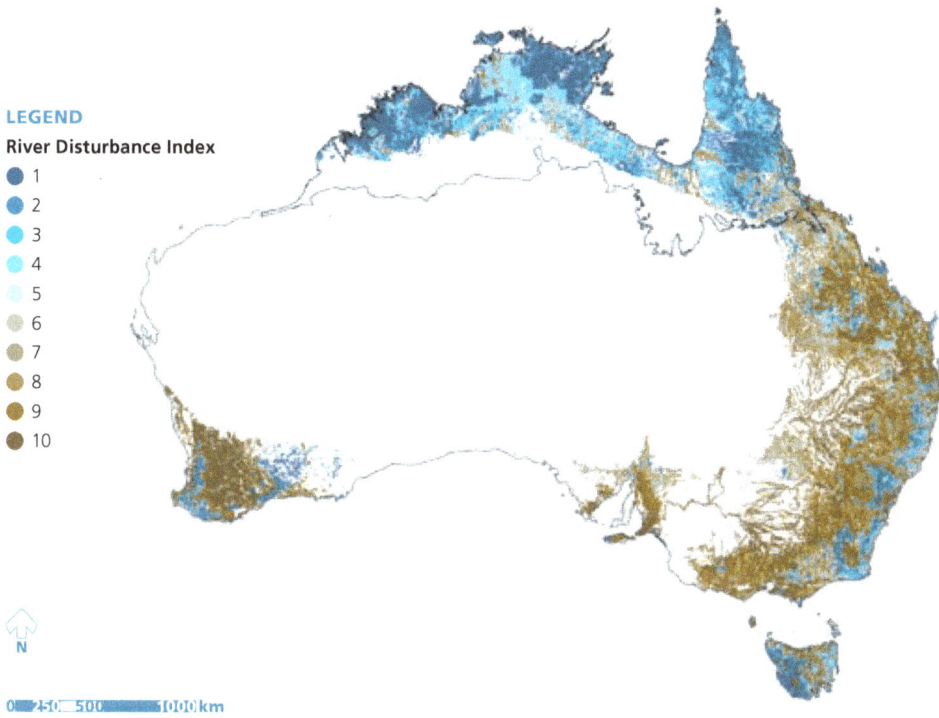

LEGEND

River Disturbance Index

- 1
- 2
- 3
- 4
- 5
- 6
- 7
- 8
- 9
- 10

N

0   250   500   1000 km

Source: Updated version of continental analysis by Janet Stein and colleagues (Stein *et al.* 2002).

① Nypa Palm Forests, Cape York Peninsula. *Photo by Kerry Trapnell*

FIGURE 4.7 **CONDITION OF ESTUARIES IN AUSTRALIA**

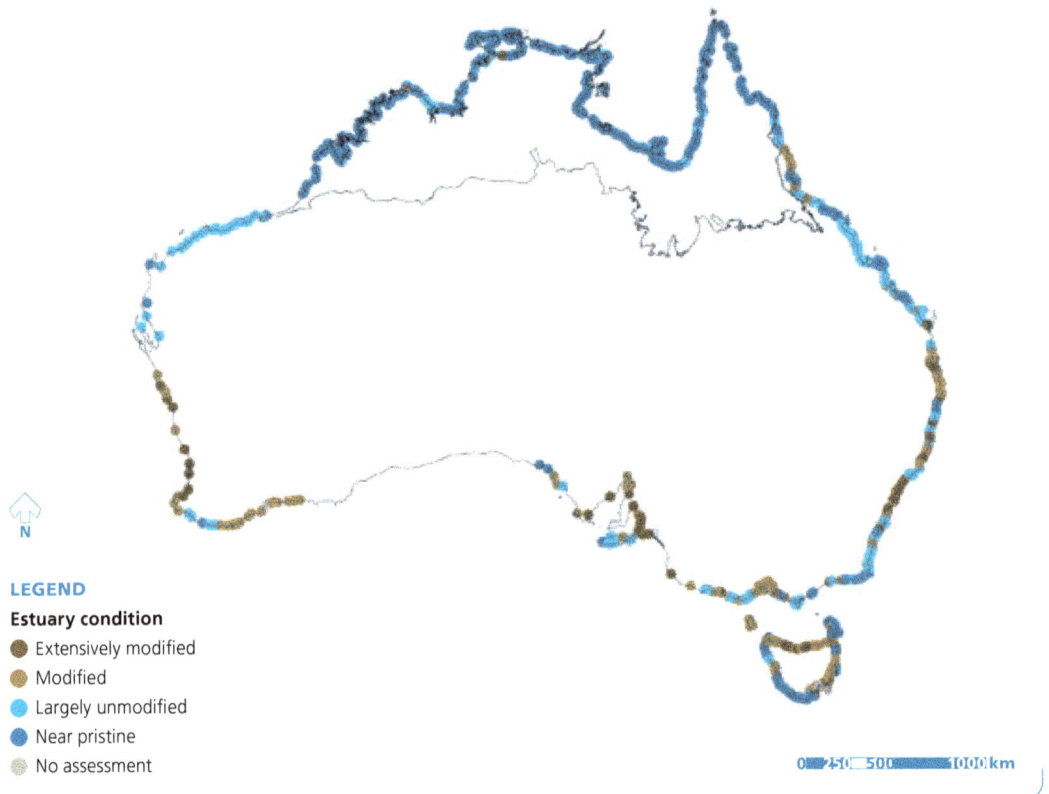

LEGEND

Estuary condition
- Extensively modified
- Modified
- Largely unmodified
- Near pristine
- No assessment

0 250 500 1000 km

Source: Land & Water Australia Audit 2004.

have systematically examined the integrity of aquatic biodiversity. Such studies (e.g. Dostine 2002) demonstrate that aquatic plant and animal communities in most Northern Australian rivers remain in largely natural condition, a feature atypical of rivers nationally or internationally.

Health of estuaries

Estuaries are coastal ecosystems where rivers meet the sea. They vary greatly in shape and form depending on the influence of river flow, tidal flow and waves. Estuaries are key habitat for a number of marine and freshwater species. The major estuaries on the coast of Northern Australia provide significant habitat for one of the densest populations of saltwater crocodiles in the world, and feeding and/or breeding areas for seabirds, waterfowl and shorebirds, dugongs and marine turtles.

Land and Water Australia recently completed an analysis of the condition of Australia's estuaries. The analysis was based on a range of biological and physical data sets and expert opinion.[3] The results from this study are

presented in Figure 4.7, which shows that most estuaries in Northern Australia are in a 'near pristine' condition, contrasting strongly with southern Australia. This is largely because estuaries in Northern Australia have a high proportion of natural vegetation cover in their catchments; minimal changes to hydrology in the catchment; no changes to tidal regime; minimal disturbance from catchment land use; minimal changes to floodplain and estuary ecology; low impact human use of the estuary; and minimal impacts from pests or weeds.

The Land and Water Australia analysis also considered the economic and social benefits that are derived regionally from these natural values, which include cultural values, pearl aquaculture, commercial (prawn and finfish) and recreational fishing, safaris, eco-tourism and traditional harvesting.

Landscapes and seascapes in good environmental condition provide invaluable environmental services to a broad range of users. One such example is in the extent to which recreational and commercial fisheries depend upon the

---

3 Details of the datasets and methodology used in the estuary condition assessments are at www.lwa.gov. au/downloads/publications_pdf/PR040674_p55-58.pdf

maintenance of productive and healthy marine environments, with this condition itself dependent upon that of the landscapes that fringe estuaries and make up the catchment as a whole (see box *'Comparing the state of Australia's woodlands'* on page 52).

THE NORTH AND AUSTRALIA'S BIODIVERSITY

A range of national analyses has examined the state of Australia's biodiversity (e.g. Land and Water Australia 2002; Australian government 2001, 2006), however the data are limited by the lack of any comprehensive, long-term and systematic monitoring programs. The component of Australia's biodiversity that has fared worst since European settlement has been the highly distinctive native mammal fauna. Of about 310 species present at the time of European settlement, 22 native mammal species are now extinct and a further ten species have had their formerly extensive continental ranges eliminated and they now occur only on a small number of offshore islands. Australian species comprise about one-third of the world's mammal species that have been lost in modern times.

The pattern of loss of Australian mammal species is notably geographically uneven (Figure 4.8). Losses – mostly of wallabies, bandicoots and larger rodents – have been most pronounced in arid and semi-arid areas, and in the more intensively developed areas of eastern, southern and south-western Australia. By far the most intact native mammal faunas are in Northern Australia and Tasmania. It is only in these areas that one can see mammal communities much as they were 200 years ago.

FIGURE 4.8 **DECLINE OF AUSTRALIA'S MAMMALS**

The numbers refer to the proportion of mammal species in each bioregion that have disappeared from more than half of that region.

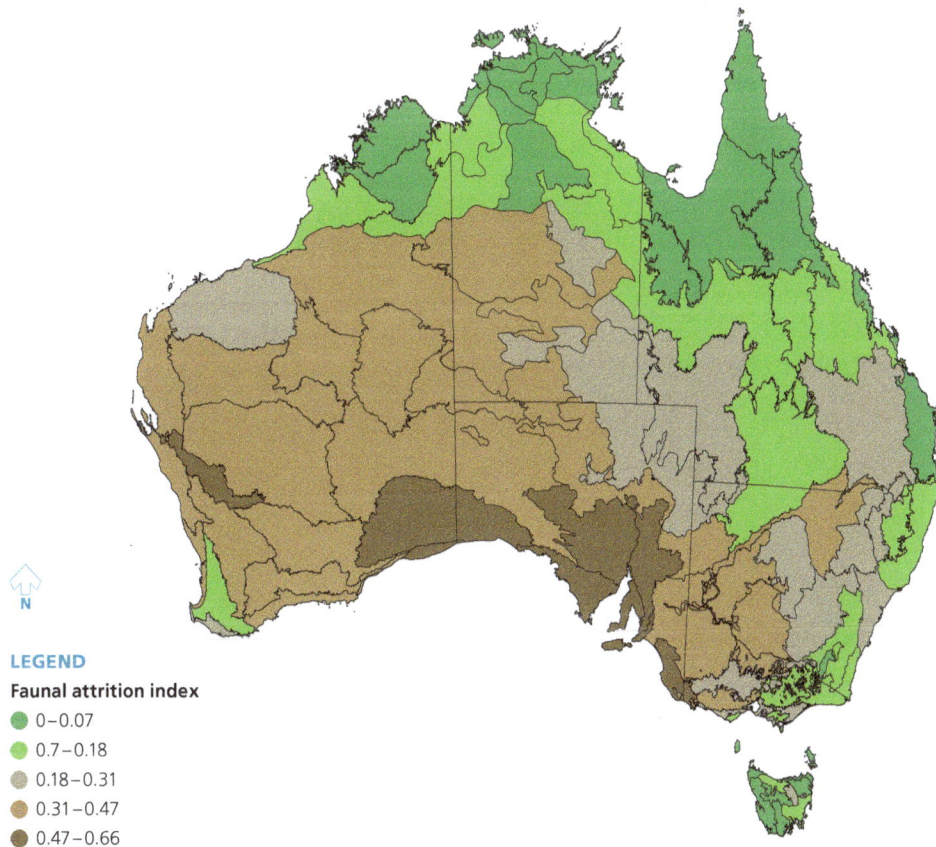

LEGEND

Faunal attrition index
- 0–0.07
- 0.7–0.18
- 0.18–0.31
- 0.31–0.47
- 0.47–0.66

Source: National Land and Water Resources Audit, Assessment of Terrestrial Biodiversity 2002 Database.

In part, these losses have been due to the spread of introduced mammals, and in particular the fox. With the recent (lamentable) introduction of foxes to Tasmania, it is only Northern Australia that has proven unsuitable for, or uninvaded by, the fox. The story is broadly similar for that other great pest, the rabbit.

There is no comparable continental analysis for other animal or plant groups, but the trends are probably broadly analogous, if less catastrophic than for mammals. A series of recent analyses have demonstrated broad-scale decline of birds in most temperate woodlands (e.g. Ford *et al.* 2001), whereas woodland bird communities remain reasonably intact in Northern Australia (Garnett and Crowley 2000; Sattler and Creighton 2002).

Another marker of the natural integrity of environments is the proportion of introduced (naturalised) plants in a region. While Northern Australia has some significant problems with introduced plants, in general the plant communities are far more intact than is typical in other parts of Australia (Table 4.2).

CONCLUSION

Australia is the only economically developed and politically stable nation containing extensive areas of tropical savanna with a high level of natural integrity. From a national perspective, Northern Australia's vegetation is also significant, as it retains savanna woodlands, rivers and estuaries with a high level of natural integrity, unlike the more environmentally degraded south. In turn these relatively intact vegetation types support plant and animal communities that are generally more intact than those elsewhere in Australia.

1 Gulf Country aerial.
Photo by Wayne Lawler

TABLE 4.2 **WEEDS IN AUSTRALIA**

Region	Number of native plant species	Number of naturalised non-native plant species	Percent of non-native (%)
Australia[a,b]	20,500	2700	12
Northern Australia			
Kimberley, species[c]	1977	108	5.2
Kimberley, including varieties[d]	2647	254	8.8
Kakadu (Alligator Rivers Region)[e]	1773	99	5.3
Top End (NT)[f]	3186	233	7.3
Set of 57 islands off north-eastern Arnhem Land[g]	665	19	2.8
Cape York Peninsula[h]	3538	274	7.2
Gulf Plains (Qld)[h]	1961	140	6.7
Comparative areas elsewhere in Australia			
Western NSW	~1514	~403	21[b]
New South Wales[j]	6539	1292	16.5
Victoria[k]	3709	1191	24.3
South-western Australia[d]	8419	1051	11
Tasmania[l]	1773	729	41

Sources: a Groves *et al.* 2003; b Cork *et al.* 2006; c Wheeler *et al.* 1992; d WA FloraBase http://florabase.calm.wa.gov.au; e Brennan 1996; f NT Herbarium, I Cowie (*pers. comm.*); g Woinarski *et al.* 2000a; h QEPA Wildnet; i Grice 2000; j PlantNet – Plant Information Network System of the Botanic Gardens Trust Version 2.0 http://plantnet.rbgsyd.nsw.gov.au; k *A Census of the Vascular Plants of Victoria*, published by the National Herbarium of Victoria, Royal Botanic Gardens, Private Bag 2000, Birdwood Avenue, South Yarra Vic 3141, Australia, www.rbg.vic.gov.au/static/viclist/viclist_ed7.pdf; l M Duretto (*pers. comm.*).

Therefore, Northern Australia stands out as a unique global asset, with nearly a third of the total area of remaining intact tropical savanna, and the largest and the most intact expanses left on Earth. Throughout the world's tropical savanna, the expansion and intensification of human populations, crops and domestic stock are all taking their toll on the natural integrity of these landscapes. The prospects for what remains are often bleak. The African countries with the largest amounts of intact tropical savanna are among the 50 poorest nations on Earth. They include Angola, the Central African Republic, Chad, Sudan and Zaire, all of which have long histories of colonial exploitation, civil war and violence. Achieving conservation goals for only a few years in such countries is fraught with uncertainty, as their people struggle to improve their economic, social, environmental and political situation, and secure their basic human rights and needs. Put simply, the quality of the natural landscapes of Northern Australia are now very rare on Earth.

In following chapters we discuss the threats to the values of the North and what is needed to mange and protect them in the long term.

PROBLEMS IN THE LANDSCAPE

In previous chapters we described some of the outstanding natural values of Northern Australia, noting that the North retains extensive natural landscapes, operating with unusually intact ecological functionality and offering exceptional levels of landscape health to its residents. It is an inspiring landscape to those who visit and a precious home to those who live there.

But not all is well in this landscape. It is not robust beyond measure. Over the last 100 or so years, almost all of its environments have been exposed to a wide range of novel pressures and impacts, and those landholders who now manage it are generally less environmentally conscious or informed than their predecessors. Even within these extensive natural landscapes, there are indications of some breakdown in ecological structure and function.

This chapter describes some of the signs of problems in this landscape and links them to possible causes. Chapter 6 provides suggestions about the maintenance, repair and planning that may be required to sustain these lands and prevent further decline.

In this chapter we argue that the values described previously cannot be taken for granted; that they will not persist through default or inaction. Without long-term and large-scale planning and without the sustained work of its land managers, the North's natural values will wither and the distinctive character of Northern Australia will be reduced to a shrinking facade. Perhaps superficially there will still be extensive savanna, but with major impairment of the underlying and fundamental ecological processes that sustain it.

The examples of loss we discuss here mostly refer to impacts on biodiversity in largely natural landscapes. We do not here consider the more acute problems that arise from vegetation clearance, the impoundment and diversion of water, and related intensive development. Such issues are considered more in Chapter 6.

WHY WORRY?

Should we worry if there are a few blemishes, some indications of imperfection in an otherwise overwhelmingly natural landscape? Who cares? Who notices? There are several answers to these questions.

KNUT DAHL & HIS BOUNTEOUS MAMMALS

One of the few accounts of the status of wildlife in Northern Australia around the time of European settlement was provided by the young Norwegian zoologist Knut Dahl, who travelled extensively in the Top End of the Northern Territory and parts of the Kimberley, from 1894 to 1896. His observations were recorded racily in the book *In Savage Australia* (Dahl 1926) and somewhat more prosaically in scientific journals (Dahl 1897). Dahl was a good and capable observer, and also respected Aboriginal knowledge of wildlife. His comments on the status of many species in the 1890s contrast sharply with what we know of their current status. For example, Dahl noted:

- For the Burrowing Bettong (in the south-east Kimberley) *'the ground was nearly everywhere and in all directions excavated by the burrows of this little Macropod … all the scrubs, and especially the slopes … are inhabited by countless numbers'*. This species is now extinct in Northern Australia.

- The Golden Bandicoot was *'very numerous in the coast country around Roebuck Bay … great numbers being brought to me'*. It now occurs in only 2–3 small areas of the Kimberley mainland, one island off Arnhem Land, and two islands off the Kimberley coast (McKenzie *et al.* 1978; McKenzie 1981; Southgate *et al.* 1996).

- For the Golden-backed Tree-rat (in the south-west Kimberley) *'the houses of settlers … are always tenanted by (this species)'*. This species now occurs only in a few small areas of the Kimberley, and there have been no confirmed Northern Territory records since 1967 (McKenzie 1981; Woinarski 2000).

- For the Brush-tailed Rabbit-rat, *Conilurus penicillatus*, *'in Arnhem Land is everywhere common in the vicinity of water'*, and *'numerous all over Arnhem Land, and in great numbers on the rivers on the lowlands'*. This species is now known from the Northern Territory mainland only on Cobourg Peninsula and one small area within Kakadu (Firth *et al.* 2006a, b).

- For the Brush-tailed Phascogale, *Phascogale tapoatafa*, *'on the rivers Mary and Katherine it was frequently observed. In fact, nearly everywhere inland it was very constant, and on a moonlight walk one would generally expect to see this little marsupial'*. This species is now rare and highly localised in Northern Australia, with fewer than ten records over the last two decades.

The species considered above provide much of the most distinctive elements of the mammal fauna of Northern Australia. In most cases, the evidence suggests widespread decline, across more developed and remote areas. Where recent evidence is available (see box 'Declines in Mammals' on page 66), this suggests that the declines may be continuing.

John Woinarski

We should heed the lessons learnt from environments that have been pushed further than the North, such as the Murray-Darling system or the intensively-grazed lands of the Burdekin system in Queensland. Once on the trajectory of decline, it becomes either very expensive or impossible to restore the health of large natural systems; and their malfunctioning has disruptive and costly impacts well beyond their immediate geographic confines. It is good economic practice to maintain environmental health over the long-term in our landscapes. It is both responsible and efficient practice to attempt to redress environmental problems as quickly as possible once they are apparent. And we do not yet know these systems well enough to figure out where the point of no return is, how far we can push the landscapes and stay within bounds of retrievability.

There are legal and policy reasons to maintain the integrity of these systems. Australia has international legal obligations to conserve biodiversity, including those stemming from the Convention on Biological Diversity. Our national and state legislations, notably the *Environment Protection and Biodiversity Conservation Act 1999*, commit Australia to conserving biodiversity and to ensuring that developments occur within sustainable limits.

Species have their own intrinsic worth. *The UN World Charter for Nature*, states that *'Every form of life is unique, warranting respect regardless of its worth to Man'*. The Charter also notes that *'Mankind is a part of nature and life depends on the uninterrupted flow of natural systems'* and that *'Lasting benefits from nature depend upon the maintenance of essential ecological processes'* (United Nations 1982). Every species has some right to persist. And the world would somehow be a poorer place without Gouldian Finches, Golden Bandicoots or any of the other species that are declining or under threat in Northern Australia.

Many species have particular values to people. Indigenous people have spiritual affiliations with many plants and animals, and their culture is degraded when these are lost. Other plants and animals may have some economic value: for example, fish are necessary for fishers; bird-watchers are an important component of the tourist industry in Northern Australia; and some unusual or currently threatened species are particular attractions. Plant or animal species may have economic value as genetic resources, for medicines, for

Indigenous arts, or for directly or indirectly supporting industries such as pastoralism.

The wildness and beauty of Northern Australia's landscapes is a fundamental attractant to the tourism industry, and its major marketing feature. Tourism is one of the region's largest economic and employment contributors, and is unlikely to flourish if the region's environments become increasingly diminished, degraded or reduced to more closely resemble those of everywhere else in the world.

We should also be concerned about the blemishes because they provide a performance measure on our land management. Most descriptions of sustainable development recognise as an explicit principle that development should entail no net biodiversity loss. Most landholders and other natural resource managers want to foster the legacy that they are responsible for and will be handing on to the next generation.

It reflects poorly on managers if species are lost or environments become degraded on their lands and under their management.

Species that are declining or environments that are degrading are indicative of problems in the landscape. They provide early warning that the landscapes are not being managed sustainably or sympathetically. They will usually be the heralds of more widespread loss. Responding to the early signs will allow us to address the problems while still solvable and will provide the most cost-effective opportunities for their solution or amelioration.

SPECIES DECLINE

Over long periods of time, species come and go, and all landscapes change. There is good fossil evidence, particularly from one of Australia's most important fossil sites, Riversleigh in the

1 *(Page 62)* Introduced Water Buffalo, Top End. *Photo by Lochman Transparencies*

2 The now extinct Thylacine can still be seen in Northern Australian rock art. The Thylacine is spoken of as the dog of the rainbow serpent and its name still survives in the Kundedjnjenghmi language. *Photo by Peter Cooke*

DECLINES IN MAMMALS

It is tantalising but inexact to compare the commentaries by Dahl and others on historic mammal status with our current assessment of their status. The apparent changes may refer to events that happened long ago and have little relevance to the conservation challenges of today. State conservation agencies, and others, are now beginning to compile increasingly precise and systematic assessments of the status of mammals (and other animals and plants) from many areas in Northern Australia (e.g. Edwards *et al.* 2003; Woinarski *et al.* 2004b, 2006b; Start *et al.* in press). This allows for more powerful and exact monitoring of changing status, for more direct assessment of the factors affecting status, and for more rapid management responses.

One of the most substantial recent assessments of the abundance of native mammals has been at CSIRO's research station at Kapalga, within Kakadu National Park. Mammals have been systematically trapped at a set of sites at this location over the period 1986–1993, and again in 1999. Over this period, trap success (the percentage of live traps that catch mammals) decreased from 24% to 3% (at one main site) and from 6% to 2% (at a series of smaller sites). Declines were significant for Fawn Antechinus, Northern Quoll, Northern Brown Bandicoot, Brush-tailed Possum, Black-footed Tree-rat and Pale Field-rat, whereas there were significant increases for only Grassland Melomys, Western Chestnut Mouse and Delicate Mouse (Woinarski *et al.* 2001; Pardon *et al* 2003). The trend in these data is for declines amongst the larger, more specialised species and for increases in a smaller set of smaller, more generalist and disturbance-favoured species. These trends are consistent with patterns of decline elsewhere in Australia, where the most extinction-prone species include larger and more specialised rodents, bandicoots, possums and quolls.

It is not yet clear how representative these data are of situations more broadly across Northern Australia (Start *et al.* in press), nor whether the picture from Kapalga is of a continuing set of trends as opposed to a more chaotic series of fluctuations without long-term direction. However, the Kapalga data represent the largest monitoring data set we have in Northern Australia. Further, they derive from the best-resourced conservation reserve in Northern Australia, so it is reasonable to assume that if there is a downward trend in biodiversity here, then it may apply at least as much in other tenures.

John Woinarski

severe climatic fluctuations between about 8000 and 80,000 years ago, coinciding also with the arrival of Aboriginal people into the Australian landscape (Johnson 2006).

Aboriginal people have left enduring images of some of these now-lost species: rock art paintings of Thylacines in the stone country of Arnhem Land are remarkably precise and vivid, beautifully representing the spirit and verve of the animal. These paintings also remind us that the Thylacine belonged to this country and was present here until a few thousand years ago.

The decline and loss of other species has been less well marked. The history of European colonisation of Northern Australia has been relatively brief, and there are few benchmarks from which to measure the changing status of species since European settlement. In the accompanying boxes, we summarise some examples, each providing insights into biodiversity change in Northern Australia over this period. The picture is not yet coherent, but there are enough indications to unsettle any cosy notions that these landscapes can look after themselves; that all species are robust in this rough and rugged land; and that, beyond the limited areas of development, there are untouched vastnesses that provide inviolable sanctuaries for all wildlife.

In Northern Australia, decline is not just some regrettable fate for a few tangential, unusual and/or ill-equipped species, it is also a pervasive feature for whole groups of species across most or all regions.

LANDSCAPE CHANGE

As with species, any assessment of the state of environments is hampered by the lack of broad-scale and systematic monitoring.

The best available information is from monsoon rainforests, particularly in the Top End of the Northern Territory. Here, rainforests occur in relatively small patches, embedded features typically associated with unusually wet (particularly springs or riparian areas) or unusually rugged (fire-protected) areas within the broader matrix of eucalypt forests and woodlands. Botanists have visited 1220 of these rainforest patches (Russell-Smith and Bowman 1992), and in each visit scored the condition of the patch with reference to four main

Gulf Country, that the biodiversity of Northern Australia has suffered some repeated major upheavals, and in the process lost many of its most spectacular plants and animals (Archer *et al.* 1991). Much of the loss of Australia's unique megafauna occurred during the

disturbance agents: fire, weeds, feral pigs and livestock (cattle and buffalo, feral or managed). The proportion of rainforest patches that they rated as 'severely disturbed' is given in Table 5.1.

About one-third of all rainforest patches were scored as severely disturbed by fire. Overall, there was relatively little difference between tenure types in the condition of rainforests. These results show that each of these disturbance factors, separately or interactively, is contributing to the degradation. The rainforest patches themselves suffer this extent of degradation, but these assessments are probably also indicative of the condition of much of the broader landscapes in which they are embedded.

An important feature of the results from this survey of rainforest patches is that the degradation is occurring through factors that

TABLE 5.1 **DISTURBANCE OF RAINFOREST PATCHES, ON LANDS OF DIFFERENT TENURE**

	Total (all tenures) (%)	**Patches on conservation lands** (%)	**Patches on pastoral lands** (%)	**Patches on Aboriginal lands** (%)
Fire	34	28	40	31
Weeds	22	29	33	13
Pigs	10	17	13	7
Stock	20	34	23	15

Figures are patches of rainforest rated as 'severely disturbed'. Percentages refer to all rainforest patches visited for each tenure type – thus, for example, 17% of rainforest patches sampled in conservation lands were severely disturbed by pigs, and 7% of the patches on Aboriginal lands were also severely disturbed by pigs. Data from Northern Territory only, from Russell-Smith and Bowman (1992).

DECLINE OF THE GRASS-SEED EATERS

Gouldian Finches, Golden-shouldered Parrots, and many other birds of great beauty have declined across the savannas over the last 100 years. Birds, such as quails, finches, parrots and pigeons, that feed largely on grass-seeds have shown major reductions in distribution and abundance over this period (Franklin 1999; Franklin et al. 2005). In some cases such as the Gouldian Finch, there have been major declines across most of their previous range.

The reasons for these declines are not immediately apparent because the habitat remains apparently intact. Across the savannas, grasses still dominate the landscapes. Grass seeds remain one of the major food sources for wildlife with massive amounts of grass seed produced annually. A square kilometre of savanna with native grasses can produce over a tonne of grass seed annually. This provides a rich, abundant and relatively reliable food source for many birds and mammals.

However, while the total amount of food remains abundant, many birds have declined. Research on some of these declining grain-eaters has shown that each species has specific needs that are not being met in many landscapes in recent years. Changed fire practices have altered habitat in subtle ways that make it harder for these species to make a living (Woinarski et al. 2005). In some areas over-grazing by stock and feral animals may also contribute to a subtle but fatal change in habitat for these birds (Crowley and Garnett 2001).

For example, Gouldian Finches face food shortages in the early Wet season after the first rains cause germination in the fallen seeds of Sorghum, their main Dry season food (Garnett and Crowley 1994; Dostine et al. 2001). They are dependent then

upon a range of mainly perennial native grass species that rarely dominate the landscape and that typically decrease in abundance in grazed situations or in contemporary fire regimes. The decline or loss of these grass species may be imperceptible to us, but lead to a fatal gap in the resource availability for finches. It is then no consolation for these finches that the increasing Sorghum may produce superabundant seeds in another part of the year. Survival depends upon eking out an existence from critical resources during periods of shortage.

Other species have variations on this problem. Changes in the savanna, some of them subtle, have led to reduced seed availability at particular times of the year, or changed the amount of cover, exposing birds to predators. These changes can reduce and eventually cause local and regional extinctions of populations of these eaters of seeds.

Barry Traill

Redhead Gouldian Finch. *Photo by Jo Heathcote*

A MEDLEY OF PROBLEMS: GOLDEN-SHOULDERED PARROT

Early European visitors to Cape York Peninsula describe a parkland landscape kept open by intense Aboriginal fire management. Sculptural termite mounds could be seen for miles through the scattered trees, and within some of them Golden-shouldered Parrots had excavated nesting hollows. But through the twentieth century this picture changed. Aboriginal burning was increasingly disrupted and cattle grazing reduced fuel loads. Thick stands of *Melaleuca* invaded once grassy drainage flats where termite mounds had been at their most dense (Neldner *et al.* 1997). At Silver Plains all burning was banned in the mid 1920s to let introduced legumes flourish for cattle feed. Within a few years the open grassland became a dense woodland.

With the grassland went the parrot – common in 1921, last seen briefly on Silver Plains after much searching in 1956. The parrots took a long time to go but research on the species suggests that the denser woodland allowed enabled predators like butcherbirds to prey more readily on nesting parrots, and gradually depleted their population. By the 1990s the species was confined to two small areas where a combination of more traditional fire management, remoteness and geology had allowed the species to persist. Even here their persistence is tenuous.

The parrot's fate is indicative of other environmental change. Black-faced Wood-swallows, open country birds with which the parrot commonly associates, are now notably less common, and Black-throated Finch also seem to be declining. The termite mounds, which may take 30 or more years before they are large enough to support a parrot nesting chamber, are also far less abundant. In some places, where parrots and mounds were common 70 years ago, neither remain. Elsewhere on Cape York rainforest is expanding. In the savannas, species that first evolved as separate species five million years ago are all but gone (Garnett and Crowley 2003).

Restoration of ecological function for these species and systems is not just a matter of declaring National Parks or removing land

Golden-shouldered Parrot on termite mound. *Photo by Cliff and Dawn Frith*

managers. Active land management is required, with restoration of fire regimes, control of feral animals and restriction of cattle movements (Crowley 2001; Crowley and Garnett 1998, 2001). Some pastoralists on Cape York have been pivotal in restoring the balance, with the result that parrot numbers have nearly stabilised where there is intensive management. Indigenous land managers in the area are also starting to undertake more traditional fire management.

Stephen Garnett and Gabriel Crowley

operate across all tenures. For these rainforest patches, conservation security is not achieved passively by inclusion within a conservation reserve. The impact of stock is not restricted only to pastoral lands. Weeds occur across all tenures. Patches close to development centres are being degraded, but so too are those remote from such development.

Beyond the limited areas of intensive modification and development, landscapes are changing across much of Northern Australia. The most noted of these changes is 'woody thickening', an increase in the density of shrubs and trees. This phenomenon has been reported widely but particularly so in pastoral areas (Lewis 2002; Fensham and Fairfax 2003;

The Brush-tailed Phascogale has declined greatly since European settlement. *Photo by Lochman Transparencies*

Banfai and Bowman 2005, 2006). There is considerable argument about its causes, with some evidence that it is related to decadal-scale climatic fluctuations, directional change in global climates and the composition of atmospheric gases, and/or to local factors related to pastoral practices (particularly reduction in fire frequency and/or intensity). The issue of woody thickening has attracted considerable comment. Many, particularly pastoralists, regard it as detrimental, largely because it reduces grass production and availability for livestock. Others see woody thickening as some balance for the extensive clearing undertaken particularly in central Queensland.

For biodiversity, woody thickening is a nuanced change: some species (or environments) will benefit and others will be disadvantaged (Tassicker *et al.* 2006). Grasslands and their associated species will be those most detrimentally affected. One such example is the Golden-shouldered Parrot, adversely affected by invasion of *Melaleuca* trees into grasslands on Cape York Peninsula (Crowley *et al.* 2004). There is also evidence of increase in the extent of rainforest patches in several regions of Northern Australia (Bowman *et al.* 2001; Russell-Smith *et al.* 2004a,b; Banfai and Bowman 2006).

Perhaps the most dynamic environment across the world is the interface between land and sea. Sea levels have changed dramatically in Northern Australia over the last 20,000 years and are likely to change substantially in the future. In many parts of Northern Australia, coastal areas are changing now. Saltwater has intruded further up river mouths, extending inland the reach of mangroves and altering the hydrological character of the highly productive coastal freshwater swamps and floodplains (Bayliss *et al.* 1998; Eliot *et al.* 1999). These habitats, important

Ord River Dam.
Photo by Stuart Blanch

for biodiversity, may be particularly vulnerable to even small rises in sea level, because they are situated in such flat and near coastal landscapes.

Some northern landscapes have lost much of their functionality and productivity. In some areas, the early days of pastoralism were highly destructive: managers were unskilled, the limits and processes of the landscape unknown, and infrastructure lacking. One of the most extreme cases of the consequent degradation was in the catchment of the Ord River. From around the 1930s to 1960s, high stocking rates, and reliance on natural water sources for watering cattle, led to a complex mosaic of environmental changes (some immediate and others still being played out). The lush riparian vegetation (often including dense stands of cane grass) was trampled or grazed out, leading to

major declines in associated wildlife (such as the Purple-crowned Fairy-wren: Smith and Johnstone 1977; Boekel 1979; Rowley 1993) and increased susceptibility of the riverbanks to collapse, and hence further habitat loss. More extensively, trampling and grazing led to loss of vegetation cover, and consequent massive erosion (and hence further loss of vegetation), especially on alluvial soils in 'frontage' country along rivers and on fertile plains of shale and volcanic-derived soils. The fertile layers of soil were lost. Rainfall became less effective as run-off increased and infiltration decreased. Small natural lakes, swamps and other water bodies – important resource-rich sites for biodiversity and Indigenous culture and hunting – were degraded or lost. 'Rangeland degradation became a self-perpetuating process' (Blandford 1979). The problem and its impacts were largely untreated until it was discovered that a direct consequence of the over-grazing was rapid siltation of the newly-built Lake Argyle, from an estimated 24 million tonnes of sediment load being deposited each year, mostly from the eroded landscapes. Subsequently, the changed hydrological conditions led to further distant changes in the ecology and physical characteristics of the Ord estuary system. The damage is still being remedied, through an expensive and long-lasting program of cattle exclusion and revegetation. But, in part, even these remedies form part of a broader problem, for almost all rehabilitation has used invasive introduced plant species, further compounding the transformation of natural systems and the patterning and process of biodiversity.

The Ord may be an extreme case (although some other catchments are directly comparable), but echoes of the same degradation are being felt across the North. For example, there are few places on Earth wilder than the upper Liverpool River in Arnhem Land, and by most measures this is land that benchmarks pristine or best condition. Here, a population of about 20 Aboriginal landowners lead largely traditional lives in an estate of a few thousand square kilometres; there are few tracks and no towns. But, in the last few decades, feral cattle, pigs and, especially, water buffalo have reached this land, and have changed its landscapes. In the fertile valleys, the trees now have distinct browse-lines, *Hyptis* and other introduced weeds form dense thickets, the kangaroo hunting-grounds are now no longer so productive, the water is no longer clear and sweet, the springs are polluted and trampled, and important cultural sites and camping grounds degraded.

THE CAUSES OF CURRENT DECLINE

Fire

The significant role of fire as an ecological process was considered in Chapter 3. As previously noted, there is an inevitability of fire in this land, but the pattern of fire regimes can provide benefit or disadvantage to different species.

Over thousands of years, the traditional fire regime routinely imposed by Aboriginal land managers shaped the landscape and the composition of its flora and fauna. The ecological equilibrium that was sustained by those traditions is now unravelling. Some species

TOO SLOW TO MATURE: HEATHLAND PLANTS

Most of the lowlands of Northern Australia are dominated by eucalypt forests and woodlands, with smaller areas of paperbark *Melaleuca* forests and woodlands, isolated patches of monsoonal rainforests, and tussock grasslands on heavier (clay) soils.

But the vegetation in the sandstone dominated uplands (and some lowland coastal sandsheets and dunes) is distinctively different. Particularly where soils in these areas are absent, skeletal or very low in nutrients, there is a very different plant community, whose floristic associations lie more with the rich heathlands of temperate eastern and south-western Australia than with that of the broader northern landscapes in which they are nestled (Specht 1981). These heathlands include endemic species of plant groups typical of heathlands in temperate Australia, such as *Banksia, Calytrix, Boronia, Grevillea* and *Drosera*, and their ecology is paced largely to a temperate rhythm (Duretto 1997; Duretto and Ladiges 1997).

One characteristic feature of heathland plants is the high proportion of species that reproduce only from seed produced by mature plants. Unless there is a durable and long-lasting soil seed bank, for such species to persist at any location, they must have a fire-free interval at least longer than that taken for a plant to grow from seed to maturity. Many of these plants require four, five or more years to reach maturity, and hence seed-production.

The problem for the northern heathlands is that the current fire regime tends to be frequent and extensive, and now does not reliably allow them the sequences of 4+ years of fire-free intervals to persist (Russell-Smith *et al*. 1998, 2002; Russell-Smith 2006). At sites where the fires recur at shorter intervals than this, populations of these plants will be locally eliminated. Where these fires are extensive, there will be far broader elimination, and re-colonisation of the plants to previously burnt sites will be unlikely.

The rich heathlands of sandstone environments in Northern Australia are now losing many of these obligate re-seeder species at an increasing number and proportion of sites. The rate of decline for many of these species is such that many are now listed nationally as vulnerable or endangered; and the richest sandstone heathland community, that of western Arnhem Land, has now been nominated (under the national *Environment Protection and Biodiversity Conservation Act*) as a threatened ecological community, the first such case for Northern Australia. The pace of life for these species has suddenly changed, and they are now fading from a world to which they are no longer adapted.

John Woinarski

GHOSTS IN THE LANDSCAPE: CYPRESS-PINE

The Northern Cypress-pine is an unusual component of the Northern Australian landscape. It is a Gondwanan relict, heavily outnumbered in the wet-dry tropics by plant species with pantropical and savanna affinities. But, in the North, it is an important tree. In the early years of European settlement, it was a highly valued resource, because of its termite-resistant and durable timber. It has many traditional uses in Aboriginal culture, mostly related to its aromatic wood and foliage.

The Northern Cypress-pine is widely distributed across Northern Australia. But across much of this broad range it is declining. This decline can be measured because, unusually for plants, dead cypress-pine trees remain standing for relatively long periods (a decade or more) and are easily recognised. There are now many parts of Northern Australia where these standing ghosts are all that remains of cypress-pine populations, unaccompanied by any new generation to replace them (Bowman and Panton 1993; Prior *et al.* 2007). It is feasible that the decline of cypress-pine may be representative of that of a range of other plant species with similar ecologies, but that these expire with far less signal.

The cypress-pine's problem is fire. Traditionally, healthy cypress-pine populations provided a very public feature of clan estates in which fire was well-managed. Cypress-pines reproduce from seed, and don't become mature enough to set seed until they are at least a decade old. The seedlings and younger trees are extremely fire-sensitive, and need a relatively long fire-free interval to be able to reach maturity. Older trees are also fire-sensitive, easily killed by severe fires, although better able to withstand less intense fires. High intensity fires will kill trees of all ages; and repeated higher intensity fires will eliminate the species locally. Persistence in a regime of lower intensity fires will depend upon the frequency of fires (Price and Bowman 1994).

The ghosts show that the cypress-pine is disappearing from much of the North, with populations persisting mostly in the most fire-protected areas (mostly highly dissected sandstone), although even in these refuges they are generally in decline.

Against this broad-scale trend, there are parts of the North where cypress-pine is increasing. These are mostly on intensively-managed pastoral properties, where fuel loads are now greatly reduced and/or where managers practise fire exclusion. These marked disparities both suggest that the landscape elements are now being re-assembled to a different equilibrium from that sustained under thousands of years of traditional Aboriginal management.

John Woinarski

and environments will benefit from the new regimes; but others will be disadvantaged. In many pastoral lands, decreased fire frequency has contributed to 'vegetation thickening'. Consequently, natural grasslands, and the plants and animals dependent upon them, have declined, because of invasion by woody plants. Across other parts of the landscape, an increase in fire frequency and/or severity has whittled the vegetation, driving out 'fire-sensitive' species, such as the Northern Cypress-pine, and diminishing ecological communities, such as sandstone heathlands. Across the entire landscape, there is now a disparity between traditional and current fire regimes, and consequently the structure and composition of vegetation is changing, in some places subtly and slowly, in others rapidly and dramatically.

Foreign animals

Many non-native animals have been introduced, deliberately or inadvertently, to Northern Australia. Most introduced animals will have some impact on the plants or animals of the country to which they have been introduced. The significance of this impact will vary between species, according to their final population size, particular ecological traits, and the extent of their interactions with native species. In this section, we consider some of the foreign animals and the mischief they bring.

The dingo was probably the first of these introductions, by Aboriginal people, about 4000 years ago. Its impacts on native Australian wildlife were probably substantial, although – now long after the event – it is difficult to tease this effect apart from other environmental factors that may have been operating at the same time. The dingo's current role is somewhat ambiguous: although it may have some detrimental impacts on populations of native animals, there is also substantial evidence that dingoes may suppress populations of the more pernicious introduced predators, the fox and cat (Pople *et al.* 2000; Newsome *et al.* 2001; Johnson *et al.* 2006). Nonetheless, dingo baiting is still widespread on pastoral lands across Northern Australia.

Cattle are now amongst the most pervasive invasive animals in Northern Australia: indeed cattle are one of the most widespread and abundant of all vertebrates in Northern Australia. The total herd size in Northern Australia is about five million: they outnumber humans in this region by a factor of about 20 to 1. The

management of about 75% of Northern Australia is now directed primarily towards improving habitat suitability for this single species, and feral cattle are present across much of the remainder of the landscape. Management activities to promote conditions for cattle include vegetation clearance, provision of artificial water sources, deliberate or consequential transformation of grass species composition (including the introduction of invasive grasses), and generally a reduction in fire frequency. Grazing itself may alter the understorey plant species composition and phenology, reduce fuel loads and hence change fire characteristics, decrease vegetative cover and hence increase run-off and proneness to erosion; and result in trampling of the tunnels and nests of ground-dwelling animals. These direct or indirect changes in environments may benefit some native plant and animal species but will also disadvantage others – perhaps many more.

Should it matter that there are some winners but also some losers? Unfortunately, it does matter – this is not simply some equitable re-balancing that emerges as a conservation neutral. There are typically more species that are disadvantaged than advantaged. Furthermore, the winners tend to be a small set of commensal species such as galahs and Crested Pigeons – those that like modified environments – and that tend to have very wide distributions. In some environments (such as Mitchell Grasslands) almost the entire landscape is devoted to

pastoralism: in such cases, there is little room left for those native species that do not prosper under pastoral land management (Fisher 2001).

Water Buffalo were also introduced as livestock, including during one of the earliest European settlements of Northern Australia, at Port Essington in Cobourg Peninsula between 1838 and 1849. At the abandonment of that settlement, much of the stock was left behind. The feral Water Buffalo thrived and spread widely across the wetlands of the Top End. By the mid-twentieth century, they had become serious pests, with population exceeding 150,000 on the floodplains between Darwin and Oenpelli (Letts *et al.* 1979). In such large numbers, feral buffalo can, and did, change ecological processes dramatically. The floodplain environments were largely denuded; and their pathways and 'swim channels' were affecting hydrological patterning. In a landscape where centimetre-scale differences in topography may affect water retention and direction of flow over tens of square kilometres, feral buffalo proved to be potent ecological drivers. The natural barriers that regulated the bounds between saltwater and freshwater were broken down and saltwater intruded into some of the most important natural freshwater floodplain systems, leading to extensive death of salt-sensitive vegetation (Whitehead *et al.* 1990). Fortunately, the feral buffalo population was greatly reduced in a broad-scale eradication program

THE OTHER SIDE OF THE FENCE: IMPACTS OF PASTORALISM

Cattle occupy much of the savanna lands of Northern Australia. They have been around for more than a century, so long that it can now be a bit hard to imagine the landscape without them. In this nominally natural landscape, the total biomass of cattle far exceeds that of any native vertebrate species: the North has become a landscape designed for cattle. For such a pervasive factor, it is surprising that there has been relatively little assessment of their environmental impact. In part, this is because of that very pervasiveness: there are now few sites in Northern Australia that can serve as cattle-free benchmarks or contrasts. Most of the National Parks, including even the 'wildest' and seemingly most 'natural', such as Kakadu, are reclaimed pastoral properties. (Indeed, a buffalo farm continues to be operated within Kakadu National Park.) Even areas with no history as pastoral properties typically support feral cattle and/or water buffalo.

However there are a few sites in Northern Australia where it is possible to measure the impacts of pastoralism on biodiversity. One such study involved a cross-fence comparison of wildlife on grazed and ungrazed lands near Townsville (Woinarski and Ash 2002; Woinarski et al. 2002). In this case, the ungrazed land was managed as a military training area, from which cattle had been excluded for 32 years. In all other respects, the environments either side of the fence were well-matched.

This study found very marked differences in the fauna between the grazed and ungrazed sides. Most of the reptile, bird and mammal species recorded showed significant differences in abundance between grazed and ungrazed lands. Some were more common in grazed lands, but these were a minority: more species were more common in ungrazed lands. Those more common in grazed lands included many commensal species– those associated with human infrastructure and disturbance. Those more common in ungrazed lands included species that feed or shelter in dense grass. Comparable results were also found at this site for two invertebrate groups considered – ants and spiders. The abundance of half of the ant species considered varied significantly between grazed and ungrazed lands, and the total richness of ants was far less in grazed lands (43 species recorded in grazed sites compared with 64 in a similar number of ungrazed sites).

This study demonstrated that, at least at this site, grazing resulted in a very major upheaval of the native fauna, and that many species were disadvantaged by pastoralism. Given the extent of pastoralism across Northern Australia, these results suggest that this industry may be having, or have had, a very pronounced effect on the pattern of fauna communities across most of Northern Australia.

John Woinarski

TABLE 5.2 ABUNDANCE OF SOME WILDLIFE SPECIES IN GRAZED & UNGRAZED LANDS

Species	Abundance in grazed lands	Abundance in ungrazed lands
Eastern Grey Kangaroo	0.13	0.63
Eastern Chestnut Mouse	0.06	0.47
Common Bronzewing	0.19	1.10
Sulphur-Crested Cockatoo	0.63	0.06
Pheasant Coucal	0.06	0.35
Weebill	1.19	4.00
Rufous Whistler	2.19	1.32
Magpie-Lark	1.06	0.28
Willie Wagtail	1.38	0.13
Torresian Crow	4.81	0.28
Gecko *Gehyra dubia*	0.88	0.03
Two-Lined Dragon	0	0.35
Skink *Carlia munda*	0.19	1.16
Skink *Ctenotus eutaenius*	0	1.35

Abundance values are based on intensive survey of 24 plots in each land use type. In each case, all differences are highly statistically significant. Note that this table includes some species that respond favourably to pastoralism and others that respond negatively.

in the late 1970s to mid-1980s, a serendipitous consequence of concerns amongst American importers about the disease status of Australian livestock (Skeat *et al.* 1996). Unfortunately, feral buffalo numbers have since increased again, especially so in more remote lands where there are few resources available for their control.

Feral pigs are now abundant and widespread in Northern Australia, especially favouring wetlands, riparian areas and rainforests, all environments associated with high conservation values. They are continuing to spread, largely unchecked: for example, in the 1990s they first colonised Arafura Swamp, one of Northern Australia's largest natural permanent wetlands, in central Arnhem Land; and in about 2000 they were (recklessly) introduced to Melville Island. Pig impacts are substantial: they have a very varied diet that includes small vertebrates, invertebrates, fruit, tubers and other plant products. They consume the eggs and young of ground-nesting birds, and marine turtles. They very efficiently dig up tuber-producing plants, including many yams that are also important components of traditional Aboriginal 'bush tucker', and they are recognised as the primary threat to several highly restricted threatened plant species. But the detrimental impact is not simply a conservation cost and problem: feral pigs have the potential to harbour many diseases, whose impacts upon the Northern Australian economy and human (and native

wildlife) health may be very significant. And pigs are notoriously difficult to eradicate.

Feral horses and donkeys occur in many areas of Northern Australia. In part because of a perceived lack of significant impact and in part because of some residual goodwill (association with stockmen history), there has been some reluctance to address control of feral horses, even in some of the most important conservation reserves in Northern Australia (Robinson *et al.* 2005). The impacts of feral donkeys may be more substantial: they can occur at very high densities, have high reproductive output, and survive relatively well even in relatively dry seasons (during which their impacts on the sparse vegetation remaining may be profound).

Feral cats are now widespread and abundant across much of Northern Australia. Although there have been few studies of their status and impacts in Northern Australia, studies from elsewhere in Australia have demonstrated that cats can have major detrimental impacts on native wildlife, and in particular on small and medium-sized terrestrial mammals. Many such mammal species are now in decline in Northern Australia, and the feral cat is a principal suspect. The persistence of many native mammal species on islands but not in nearby mainland areas (Johnson and Kerle 1991; Abbott and Burbidge 1995; Southgate *et al.* 1996) provides

1 Feral horses cause significant problems through overgrazing in many parts of the North. *Photo by Charles Tambiah*

2 Overgrazing by feral donkeys has been a major problem in the Kimberley, until recent control efforts by the West Australian government. *Photo by Charles Tambiah*

75

some evidence for the destructive impacts of feral cats in Northern Australia generally.

Parts of Northern Australia have been invaded by the introduced House Mouse and Black Rat. While most populations of these rodents remain associated with human infrastructure and more intensive development, some populations have established in less modified natural environments, including on some remote Northern Australian islands.

In contrast to temperate Australia, and to most other parts of the world, there are few foreign birds in Northern Australia, and even these are currently highly localised. However, in recent years the House Sparrow is spreading across more of the north, and there has been an increase in cases of foreign birds reaching Northern Australia on ships and other transport (e.g. Chapman 2000).

The cane toad is perhaps the most notorious invasive animal in Northern Australia. Its history is now well documented; and there is much public scrutiny of its continuing spread across Northern Australia. Many native predators, including some snakes, goannas and the Northern Quoll, fall victim to its toxin (Burnett 1997; Doody *et al.* 2006; Smith and Phillips 2006). However, it is not yet clear whether or not these species will gradually recover, and its less conspicuous impacts, on its invertebrate prey or native frog competitors, remain largely unknown. We are living in its unfolding story, as one of the world's most successful and adaptable creatures discovers a land particularly to its liking.

Less immediately obvious is the invasion of Northern Australia by a series of foreign invertebrates. For most people, an ant is an ant, and introduced ants are indistinguishable from natives. But Northern Australia is being invaded

by a group of foreign 'supertramps', aggressive colonial species, that can wreak astonishing damage to native plant and animal communities (and to agricultural and other land uses). These ants include the Yellow Crazy Ant *Anoplolepis gracilipes*, which has invaded parts of eastern Arnhem Land, and the African Big-headed Ant *Pheidole megacephala*, which has invaded rainforest patches, urban and peri-urban areas at many sites in the Northern Territory (at least). Because of their aggressive nature and large population sizes, these invasive ants typically substantially reduce the diversity of native invertebrates, and may either directly or indirectly alter ecological processes, with detrimental impacts on native vertebrates, plants and ecosystem health (e.g. Hoffmann *et al.* 1999). A recent survey of the largely undeveloped Tiwi Islands revealed the presence of nine introduced ant species (Andersen *et al.* 2004; Hoffman *pers. comm.*). These invaders have slipped under our radar.

The European Honey Bee, *Apis mellifera,* has been deliberately introduced to many parts of Northern Australia, to provide honey and to increase the pollination rates in some horticultural crops. It has since spread widely, invading natural areas well away from its points of introduction. Like other introductions, its spread causes a rent in the fabric of the natural ecological systems. Much of the environmental cost of the introduction of honey bees is out of sight: their gradual but inexorable spread to natural areas, their usurpation of hollows (that otherwise would be used by a wide range of native vertebrates and native bees), and the competition they present to native species at nectar and pollen sources. Given wild honey produced by stingless native bees is an important bush tucker resource for Aboriginal people, this spread may have wide-ranging detrimental consequences, entirely unconsidered and discounted by those that introduced, and approved the introduction of, this invasive species.

Marine and aquatic systems have also been invaded by foreign animals. Some river systems in Northern Australia now contain African and South American species, Nile Perch and Mosquito-fish respectively, and both are increasing in range. The South American Black-striped Mussel invaded Darwin harbour from a foreign ship in 1998, but a rapid response eliminated it. This species is a major threat to marine ecosystems, and the economic costs (to marine transport and fisheries) of any invasion would be in millions of dollars per year.

We've brought many other introduced species – cockroaches, fleas, worms, lice, and others – to Northern Australia, largely unnoticed. There is no tally of such species, and there has been no attempt to investigate the extent to which they may have spread to natural areas, and have had detrimental impacts upon biodiversity. Mostly, we remain blithely ignorant of their presence and detrimental potential, and largely uncaring. But we subvert this landscape not only through conspicuous major modifications, but also with the accumulation of almost imperceptible small changes, and especially so where these changes ripple out from their point of origin.

Disease

One more recognised example of the risks posed by the spread of invasive organisms is the recent substantial concern associated with the possibility of spread of bird flu to Australia, through migration of infected birds from Indonesia or elsewhere in Asia. While this spread is unlikely, it does provide an illustration of the potential major consequences, in this case to both wildlife and humans, of the spread of foreign diseases. It also reminds us that Northern Australia may well be the open frontier for such diseases. Some of these diseases, such as Japanese encephalitis, are likely to reach Australia 'naturally', without direct transport by humans. Others have come, or will come, as undesired and/or undetected baggage. It is likely that many of these newly introduced diseases will affect native plants and animals in Northern Australia. For example, dieback, probably caused by invasive *Phytopthera*, has recently been reported in some eucalypt stands in Northern Australia. In this case, detection is made easier by the highly conspicuous symptoms: but in most cases, we will remain

HABITAT LOSS & THE EMERGENCE OF INFECTIOUS DISEASES IN FLYING-FOXES

Over the past 12 years three new viruses have emerged from flying-foxes into domestic animals and humans in Australia: Hendra virus, Menangle virus and Australian bat lyssavirus. It is highly unusual for so many viruses to emerge from one host in such a short period of time and signifies a significant change in host ecology.

Taking a closer look at flying-foxes one may view the raucous large colonies in our urban centres and believe that their populations are robust or even increasing. This common misconception could not be further from the truth. In fact the presence of flying-foxes in urban areas is a symptom of problems that seriously threaten most flying-fox populations in Australia: habitat loss and fragmentation.

Flying-foxes move nomadically across the landscape tracking irregular flowering and fruiting events in our native forests. As forests are cleared and their native food supply dwindles, flying-foxes have begun to seek alternative food sources such as the native and exotic flowering and fruiting trees in urban areas. These regularly watered urban food resources are reliable and available year round, reducing the motivation for long-distance energy expensive foraging expeditions. Consequently, urban flying-fox camps are growing while more rural locations are being abandoned. This change in flying-fox ecology has major consequences for both ecosystem health and human health.

As flying-foxes move away from their natural habitat, they deprive our native forests of their essential pollination and seed dispersal services and come into closer association with human habitats, increasing the risk of emergence of rare viruses. Intact habitat is critical to maintain the ecology of flying-foxes, the health of the forests they pollinate and the health of humans and their domestic animals.

Raina Plowright

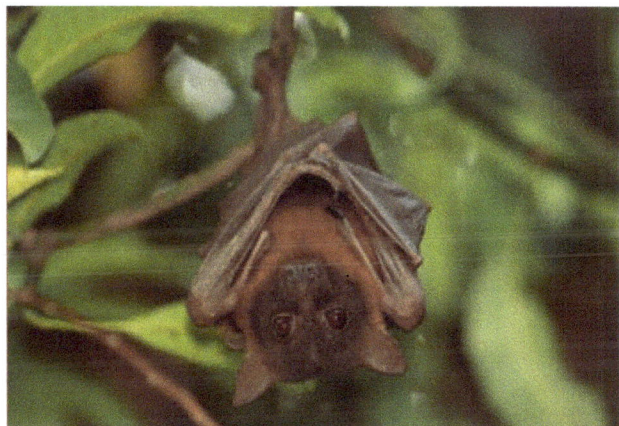

Little Red Flying-fox. *Photo by Raina Plowright*

(or have remained) largely unaware of the spread of diseases to native plants and animals.

Invasive plants

Native vegetation dominates almost all landscapes in Northern Australia; and the vast extent of the landscape appears natural. But there are conspicuous exceptions, and there is a gradually increasing proportion of the flora that is alien, and increasing abundance and distribution of many introduced plants. On current trends, much of the landscape will be transformed by invasive plants.

For most regions of Northern Australia, naturalised (that is, now wild-living) foreign plants comprise about 5% of the total flora. This is relatively low by Australian standards (where the overall proportion is about 12%), but still substantial. Invasive plants are diverse and widespread even in remote largely unmodified lands, and in National Parks: for example Kakadu contains about 100 species of alien plants in a total flora of about 1800 plant species (Cowie and Werner 1993; Brennan 1996).

There are widely acknowledged problems with some of these invasive plants. Those that are universally recognised as weeds include the floodplain shrub Mimosa *Mimosa pigra* (from tropical America), the floating aquatic weeds *Salvinia molesta* (from South America), cabomba *Cabomba caroliniana* (from tropical America) and water hyacinth *Eichhornia crassipes* (from tropical America); the invasive woody shrubs *Parkinsonia aculeata* (from central and South America), *Acacia nilotica* (from tropical Africa and Asia), rubber bush *Calotropis procera* (from Asia), chinee apple *Ziziphys mauritiana* (from tropical Africa and Asia), parthenium *Parthenium hysterophorous* (from tropical America) and hyptis *Hyptis suaveolens* (from Asia).

This smorgasbord of some of the world's most harmful weeds combines to degrade values of lands and waters for all land users, including for pastoralism, horticulture, recreational fishing and Indigenous use. Removed from their normal controlling influences, these species have prospered and spread through large areas of Northern Australia, in some cases completely dominating particular environments or localities (e.g. Cook *et al.* 1996). These species turn Australian environments into a medley of odd global elements, generally substantially less suitable for the native plants and animals of Northern Australia. But in part because of their disadvantage to all land users, these species are now the focus of a range of control and weed management actions, sometimes successful, sometimes forlorn, and usually expensive. For example, Kakadu National Park has employed a team of three to four staff full-time for at least the last two decades mostly to attempt to keep Mimosa out of the park, with an annual cost of about $500,000. While this has been largely successful to date, it does not solve the core problem: Mimosa is still largely unchecked outside the park, and will invade the park should the intensive exclusion management be halted.

In part because the plants considered above are conspicuous and in part because they bring problems to all land users, weed management has focused on those species. But these are just one element of the environmental weed problem of Northern Australia (Grice 2006; Martin *et al.* 2006). Probably more pernicious is the continuing spread, including ongoing deliberate planting, of introduced invasive pasture grasses (and other pasture plants) largely for perceived increase in pastoral productivity, but also in some cases for rehabilitation and erosion control. In a recent article, Cook and Dias (2006) charted the history of introduction of these (mostly African) plants to Australia, particularly in the north. They note that the spread of these plants was a calculated and deliberate strategy to transform the full extent of Northern Australia from landscapes of native vegetation to designed landscapes dominated by 'more productive' foreign plants that would form the foundation for a greatly increased cattle industry. They note that the number of foreign grass and legume species deliberately introduced to Australia was about twice the number of native species present in Australia, and represented about 20% of the total world's species of grasses and legumes.

This was environmental transformation planned at a massive scale, explicitly with little or no regard to environmental consequences or the values of land uses other than pastoralism. It was more subtle than working with bulldozers, but its environmental consequences have probably been far more pervasive. The itch to de-Australianise these landscapes came early, from some of the first pastoral enterprises, but peaked especially in the 1950s, when teams of agricultural scientists worked assiduously to import more and more rampant species.

The philosophy and practice continues today. There is an ongoing push for pastoral intensification, and the principal mechanism for this intensification is through the deliberate and authorised spread of foreign invasive grasses, including Buffel Grass, Gamba Grass and Para Grass. It is a peculiar thing, but most of the pasture plants introduced are found, often too late for effective remedial control, to be useless for pastoral purposes, and in many cases actually become pests even for pastoralism. There is a recurring theme in the history of pastoral introductions for plants to be proselytised enthusiastically, planted widely, nurtured and tended, found to be useless, become uncontrollable and then, usually 20–30 years later (typically when it is someone else's problem to deal with) become officially declared as weeds.

Gamba and Para Grass are moving towards the end-game of this process now, but it is an end point marked by almost irretrievable loss. Once these are established in the landscape they can't

1 Monitoring rangeland condition, Mornington Station, Kimberley (Jo Heathcote, Emma Flaxman). *Photo by John Augusteyn*

2 Gamba Grass, introduced pasture grass that is now a major weed in the Northern Territory. *Photo by Stuart Blanch*

79

1. Gubara, Kakadu National Park, Top End. *Photo by Glenn Walker*

2. Simon Ward and Graham Friday, a member of the Lianthawirriyarra Sea Ranger Unit, doing wildlife survey work on Sir Edward Pellew Islands, near Borroloola. *Photo by Felicity Chapman*

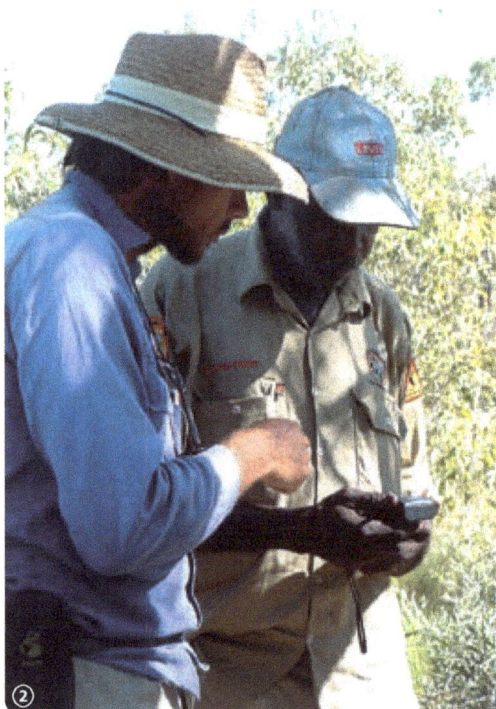

be easily whistled back. The very qualities that were sought in a foreign plant by the pastoral interests – good reproductive potential, ability to spread from point of introduction, hardiness, ability to out-compete native species, high production – are the very same qualities that make for weeds (Whitehead 1999). But, while there is still some hope that they may bring benefit to pastoralism, there has been an enduring reluctance to regulate or prohibit their use, or to officially recognise them as weeds.

The problem is not widely recognised because for most people a grassland or grassy understorey looks much the same whether dominated by invasive or native grasses. Furthermore, pastoralists and land managers may categorise a monoculture of invasive foreign grasses as a healthy landscape, a valuation based on a very narrow perspective or understanding of ecological function and process, and of what is meant by environmental sustainability.

What do these plants do that is wrong?

Firstly, they displace native plants. Plant species richness in landscapes dominated by invasive grasses is almost invariably lower than in landscapes without them (Fairfax and Fensham 2000). In some cases, the native plants displaced provide key resources for fauna, and so inexorably these animals will also become rarer or be lost from invasive grasslands. Invasive pasture grasses typically also produce more growth – for this is the key attribute for which they were selected. In most places (other than where there may be very high densities of cattle) this results in a substantially increased fuel load and grass cover. This strongly disadvantages the many native plants and animals that require some patchily open understories to germinate, grow or forage. It also means that fires will burn at greatly increased intensities, typically being more destructive, more extensive and less patchy. For Gamba Grass, fuel loads are typically four to five times that of native grasses, and fires eight times the intensity (Rossiter *et al.* 2003). Scientific simulation modelling has shown that over a period of a few decades, Gamba Grass infestation will transform a native savanna woodland to an African-style grassland, with wholesale elimination of trees. Fire-sensitive plants, animals and communities will be lost. Modelling has also indicated that all or most regions of Northern Australia have viable habitat for the current medley of invasive pasture plants; and indeed the current spread of many of these invasive pasture grasses is proving to be rapid.

The vision of the landscape engineers of the 1950s is not far over the horizon. But it may be a dystopic reality, as the alien grasses and their consequent fires become uncontrollable and result in increasing damage not only to environmental values but also to infrastructure, health (e.g. through increased incidence of asthma: Johnston *et al.* 2002) and public safety.

The invasive grasses do not respect tenure boundaries, and will increasingly spread from pastoral properties to lands managed primarily for biodiversity conservation. This will erode the functionality of those conservation areas, greatly increase their management costs, and further smear the landscape to a continuum of modified and degraded environments. A perverse outcome of the cross-boundary spread of invasive pasture grasses is that some National Park managers (and their pastoral neighbours) are now contemplating the need to agist cattle in conservation reserves in order to keep down the fuel load of unwanted invasive grasses.

DEPOPULATION AND LACK OF RESOURCES

There was a time when conservationists idealised land devoid of humans and their artefacts as the ultimate standard for biodiversity protection. That vision does not work in Northern Australia. In Northern Australia, most of the remote wild lands are the estates and homelands of Indigenous people, and their ancestors have shaped and protected those landscapes for millennia. In Northern Australia, biodiversity is detrimentally affected by factors – inappropriate fire regimes and the spread of introduced plants and animals – that require active human intervention. A depopulated 'wilderness' may become barren (Bowman *et al.* 2001; Yibarbuk *et al.* 2001), the land losing its wildlife and its people their culture.

But having people in the landscape does not alone make for good management. Communities blighted by poverty, unemployment, poor health and poor education may see management for biodiversity as an unaffordable luxury, and look to their land not so much for its nurture but rather as a resource to be sold.

In many Indigenous lands, there are now few people living on country, and few resources available for land management. Some of the most expert land managers in Northern Australia look after their country full-time and unpaid, but even this expertise may be lacking when faced with novel threats and breakdown of land management in neighbouring estates. The Arnhem Land plateau provides a simple illustration of the issue (Figure 5.1). An artificial tenure boundary splits this plateau, with Kakadu National Park to the West and Arnhem Land Aboriginal land to the east. Kakadu is a relatively well-resourced National Park and annually spends about $725/km^2 on land management (a figure about par with world standards: du Toit *et al.* 2003). But cross the line to the east, and the amount spent on land management is less than $1/km^2, and the number of people living in the plateau landscape and available to do that management is fewer than one person per 100 km^2. That land has been largely depopulated and its threatening processes left mostly unchecked. The problem is not constrained: because most threatening

FIGURE 5.1 **FUNDS SPENT IN KAKADU NATIONAL PARK & ADJACENT ABORIGINAL LANDS**

Annual funding per square kilometre in adjacent land tenures in the Top End. The different land tenures have similar environmental values.

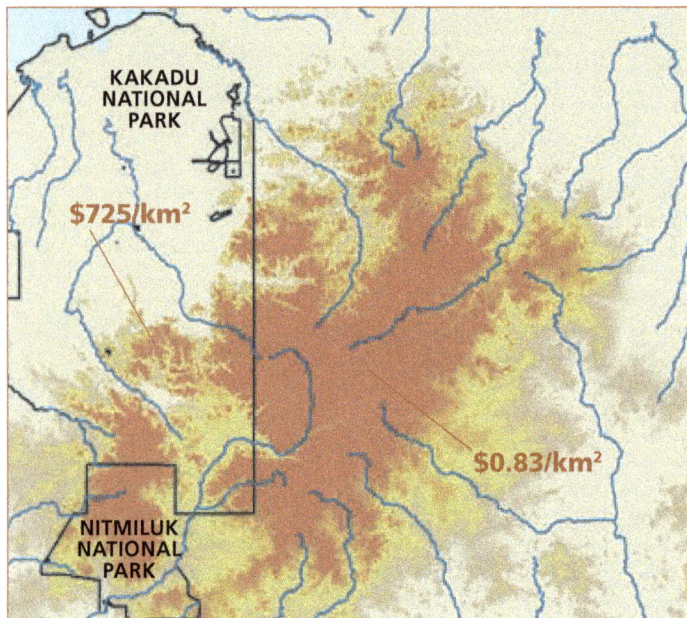

by increased levels of atmospheric carbon. In Northern Australia, there is increasing evidence of seawater intrusion to coastal floodplains and wetlands, and spread of mangroves further up coastal waterways (Eliot *et al.* 1999).

These direct environmental changes are likely to continue and perhaps accelerate over the next few decades. Most susceptible will be the internationally-significant wetlands, particularly the extremely fertile and biodiversity-rich floodplains, as even small rises in sea levels will have extensive impacts on this low-lying flat landscape. Although climate change predictions for Northern Australia are very imprecise, it is likely that the main features will be (i) minor rise in temperatures; (ii) increased rainfall; and (iii) increased incidence of severe storm events (most notably, cyclones). A possible significant environmental impact of these changes may be an increase in fire intensity, frequency and extent, if fuel levels increase following Wet seasons with high rainfalls. The changes in rainfall and temperatures may also differentially affect survival and habitat suitability for some native plant and animal species, perhaps especially crocodiles and some turtles, where sex determination in eggs is geared to a narrowly defined temperature trigger point. Climate change may also affect those species in Northern Australia with restricted geographic range and/or narrow climatic tolerances, such as those species occurring in mountain tops in north-eastern Australia. These projected climate changes may also increase the competitive advantage of some foreign species and the incidence of some infectious diseases, as these may be more attuned to the Northern Australian climate of the future.

However, in general, the direct impact of climate change on the environments of Northern Australia are likely to be subdued relative to those in other parts of the world, because (i) the projected climate changes are relatively minor; (ii) the environments remain highly connected and extensive (permitting some possibility of movements to track suitable climate); and (iii) there are no abrupt and pronounced climatic specialised features in Northern Australia (such as alpine areas) that elsewhere are most vulnerable to climate changes.

Instead, the impacts of climate change on the conservation values of Northern Australia are likely to be mediated more by indirect (extrinsic) factors. The future is likely to bring greater pressure on Northern Australia for production,

processes occur pervasively across tenures. Unmanaged lands then provide the ongoing genesis for feral animals, weeds or fires to spread back to more intensively-managed lands.

Limited resourcing is not an issue that affects Indigenous lands only. In Northern Australia, most National Parks have appreciably fewer management resources than Kakadu; and rising running costs and reduced profitability constrain the resources available on pastoral lands for land management activities other than those most focused on short-term pastoral productivity.

CLIMATE CHANGE

The environments of Northern Australia will be affected directly and indirectly by rapid global climate change. Indeed, some effects may already be evident. In north-western Australia, rainfall has increased substantially over the last 100 years, particularly in the last 30 or so years. In many regions, the density of trees has increased, woodlands have extended into grasslands, and rainforests into woodlands and open forests. This change is comparable to changes observed in savanna regions elsewhere in the world, and probably driven at least partly

if the prime horticultural areas of temperate Australia, and perhaps even more importantly, much of south Asia, become increasingly unproductive in the future as a result of a critical decline in rainfall and water availability.

SOLACE

This is a chapter that views the glass half empty; that presents a series of examples of decline and degradation. We chronicle these cases not to downplay the natural values of Northern Australia, nor the current management of that land, but rather because we should recognise and heed the early warning signs, and not complacently think that the extent and relative naturalness of the landscape will be adequate protection for its biodiversity.

The natural values of Northern Australia are substantial. But their persistence cannot be taken for granted. Some important components of biodiversity are faring poorly under current management practices, policies and levels of resourcing. The following chapter offers some hope for better sustaining the North's natural values.

1 Above Gunlom Falls, Kakadu National Park, Top End. Adjacent Aboriginal Lands receive much less management resources. *Photo by Glenn Walker*

A SUSTAINABLE FUTURE FOR NORTHERN AUSTRALIA

PROTECTING THE NATURAL LEGACY

In part through the good management of the North by many of its landowners, in part because historical attempts at intensive development have been limited and fitful, and in part because of some resilience in the landscape itself, we have been gifted a great natural legacy – the largest intact savanna remaining on Earth, an extraordinarily vast, natural landscape with a rich biodiversity of international significance. The challenge is to ensure that this legacy is appreciated, respected and retained.

This legacy provides an unusual opportunity to live in and use an extensive landscape, sensibly and sustainably. We can be guided by the benefits of good knowledge of the ecological limitations of this land and by the lessons from elsewhere of the consequences of unsustainable development. There are few comparable opportunities in the world.

We take as our starting point that Australians do not want to repeat the environmental mistakes of the South, but rather want to maintain the North's natural legacy, while also enabling appropriate economic development to meet

communities' needs and aspirations. This chapter considers how such a future may be achieved.

CONSERVATION PLANNING

Historically, the environmental values of Northern Australia have been discounted, and the environmental impacts of economic activities in Northern Australia generally overlooked. The rationale occasionally put forward has been that with so much land, water and bush there is simply no need to be concerned about any local impacts. More recently, it has become mandatory to assess the local environmental impact of large site-specific projects such as mining. However, the process frequently ignores cumulative environmental impacts outside the immediate local district. Similarly, the widespread and pervasive impacts from activities such as pastoralism are rarely assessed, and the cumulative effects of many small changes to the environment in a region are not considered. Instead, it is assumed that there is abundant land to cater for natural values somewhere else.

These are symptoms of a reactive approach to land use and conservation planning; an approach

characterised by ongoing episodes of narrowly focused considerations, usually in response to specific acute development pressures. Here, we argue that this is not strategic, and will result in the incremental erosion of conservation values. Instead, we propose that a sustainable future for Northern Australia can be achieved only through broad-scale and long-term planning, in which primacy is given to the national and international significance of the natural landscapes of Northern Australia.

The conventional target-based approach to conservation planning

Elsewhere in Australia, and overseas, conservation planning has largely focused on a land allocation process, where a portion of the available land parcels is set aside as National Park, largely surrounded by lands not managed for conservation.

This conventional approach typically focuses on achieving, as efficiently as possible, a target level of protection through reserves, National Parks or other protected areas. This target is based on the reservation of representative samples of each class of ecosystem that can be mapped (in practice, usually defined in terms of major vegetation types); critical habitat for, or

populations of, particular target species (typically threatened plants and animals); and other specified special features. Conservation targets are usually expressed as a nominated percentage level; for example, many planning exercises seek to protect within conservation reserves 10–30% of the historic extent of every vegetation type. The regional conservation objective is that this set of conservation lands comprises a comprehensive reserve network. This approach has been advocated in parts of Northern Australia (e.g. Burbidge *et al.* 1991; Price *et al.* 2000).

Assumptions underlying this approach are that:
- Target levels are scientifically robust, and reservation to that target level will provide long-term security for the represented biodiversity;
- The environmental layer used in target-setting provides a good surrogate for the distribution of most elements of biodiversity;
- A relatively small network of protected areas can ensure the long-term conservation of biodiversity at all levels (genetic, species, ecosystem), and will be reasonably insulated from the impacts of the surrounding, majority land uses;
- Natural values can be traded-off against competing land uses by identifying alternative locations (e.g. habitat loss at one location can be compensated by protecting habitat at another location); and
- There are enough land parcels available to set aside within conservation reserves the full suite of environments.

Why the conventional approach will fail in the North

Unfortunately, these assumptions are often invalid, and the approach has major problems, most evident in largely natural intact landscapes. Land use allocation exercises may rapidly descend into arguments over site-specific issues – the demonstrable value of a particular site considered in isolation. In such debate, the currency of conservation values is often difficult to match against that of an explicit or perceived economic value. The end-result and typical aim of this approach is usually a landscape dissected by contrasting land uses, with isolated conservation areas scattered throughout a landscape largely transformed by modern land use activities. Generally, these conservation areas are too few in size and number, and are in areas of low value to other land users (often because these are infertile, too steep, or otherwise of low productivity).

③

Further, the approach was designed to wring some conservation outcome within regions that typically had been largely fragmented and disturbed, where dedication to conservation reserves of some of the remaining patches of native vegetation was the least worst of the limited conservation options possible. For the intact landscapes of Northern Australia, the conventional approach of focusing conservation effort towards establishment of National Parks representing 10–30% of the extent of every vegetation type implies that the dominant 70–90% of the landscape may lose its conservation attributes.

Elsewhere in Australia (and in some locations in Northern Australia) the failure of conservation planning to take into account the ecological processes and connections that make the landscape work and keep it healthy has caused major environmental problems. The environmentally harmful effects of

unsustainable land management in the Murray-Darling basin and the south-west of Western Australia are well documented. Failure to protect hydro-ecological processes has caused salinity and degraded rivers; as a result, there are major salinity issues on farmland, and many nomadic and migratory species have greatly declined due to the removal and degradation of key habitats.

In most cases, the degradation of ecological processes is due to the accumulated impact of incremental changes that alter the country (e.g. another paddock sown to exotic pasture grasses, a few more megalitres diverted for irrigation, another patch of tree clearing), repeated tens or hundreds of times over a region.

Although an extensive network of protected areas is a necessary component of the long-term maintenance of the natural values of Northern Australia, it will not be sufficient for their protection. In particular, a conventional approach that focuses on just achieving target levels of protection will not provide a sustainable future for Northern Australia, because:

1. It fails to incorporate the landscape-wide ecological processes that sustain natural values. For example, however large the protected area network may be, many critical natural values will depend for part of the time on lands and resources outside conservation reserves. Ultimately, species and other natural values within isolated conservation lands will decline and degrade as the ecological connections that they depend upon break down.

2. This approach does not recognise nor maintain the greatest conservation asset of Northern Australia – its extensive landscape connectedness and intactness.

3. There is little unallocated land available. In contrast to conservation planning in most temperate regions of Australia, where new conservation reserves can be created from 'vacant' crown land or lands long established as forest reserves, there are existing proprietal interests in almost all lands in Northern Australia.

An alternative approach – managing across the landscape

A new approach to conservation, land management and land use has been developed in sparsely-populated and largely natural extensive landscapes in North America (Soulé and Terborgh 1999), and subsequently re-tuned for Australian conditions (Soulé *et al.* 2002; Mackey *et al.* 2007). This approach is based on the maintenance of ecological processes and emphasises the maintenance, or in some cases restoration, of large-scale connectivity of natural ecosystems. It advocates the development of collaborative partnerships for land management across landholders and promotes the establishment of significant conservation reserves as the core areas of an interconnected conservation network. And, the approach supports the development of a conservation economy, including livelihoods associated with land management.

Within Australia, a major government review recently advocated such large-scale collaborative conservation planning as a necessary response to attempt to reverse ongoing biodiversity decline (Natural Resources Policies and Programs Committee

Biodiversity Decline Working Group 2005). It is far more feasible to attempt to implement such an approach in Northern Australia than almost anywhere else in the nation.

The likely impact of global climate change further emphasises the need for conservation to be based on large-scale connectivity, because smaller and more isolated conservation reserves can be expected to come under increasing pressure.

Here, we draw upon these initiatives and propose that the sustainable future of Northern Australia must be based on inverting the conventional planning approach of leaving to conservation the residues or samples left over after other land uses have taken their priority picks (Nix 2004). Instead, we propose that maintenance of the natural landscapes, with their full suite of natural values and processes, should explicitly underpin the sustainable future of Northern Australia.

From this premise, we derive a series of guiding principles for Northern Australia:

1. *The natural environments must be valued recognising their national and international significance;*

2. *The ecological integrity of the processes that support life must be maintained;*

3. *The population viability of all native species must be protected;*

4. *Thresholds defined by the limits to ecological integrity, including cumulative impacts, must be used to assess and guide development options; and*

5. *The contributions of all property-holders and managers are needed to maintain the North's natural values.*

It follows from these principles that connectivity of natural vegetation and waterways must be maintained at property and regional scales. Biodiversity and links of interconnecting ecological processes occur on all lands and waters, and all properties, in Northern Australia. All property-holders and managers have the opportunity to enhance or degrade the natural values on their properties, and thus on the properties and landscapes around them. Society should expect a reasonable level of sustainable management by property-holders, and beyond that provide incentives for proactive environmental management.

LIVING IN THE LAND

The sustainability and conservation planning approach described here is not about locking up land, keeping people out, stopping economic development, nor keeping Aboriginal landholders in stasis. For a range of compelling social and economic reasons, the North requires development, and the conservation management of Northern Australia will require substantial funding. However, if the natural values of the North are to be retained, then that development must fit the nature of the country.

To ensure the long-term protection of the North's natural values, priority should be to develop and improve the economy in ways that are compatible with the North's environment. In particular, further support is needed for developing the 'conservation economy', that is, income-generating economic activities that provide employment opportunities and benefit the environment, such as management of parks and control of invasive plants and animals. These kinds of natural resource management activities provide benefits for all Australians – ensuring a natural base for northern tourism and for commercial, recreational and Indigenous fisheries. Indeed, such activities are already a significant part of some regional economies in the North. In some communities, the conservation economy may also provide direct social benefit through alleviating underemployment, through maintaining cultural traditions and strength, and through improving health. Such social benefits will serve to improve outcomes from, and reduce economic costs associated with, the current delivery of health and welfare services. The benefits of such conservation management investments on Indigenous lands is increasingly being realised, and funded, by national and state governments (e.g. $48 million has been allocated in the 2007 national budget over four years for the *Working on Country* program), and the Indigenous Protected Areas program is recognised as

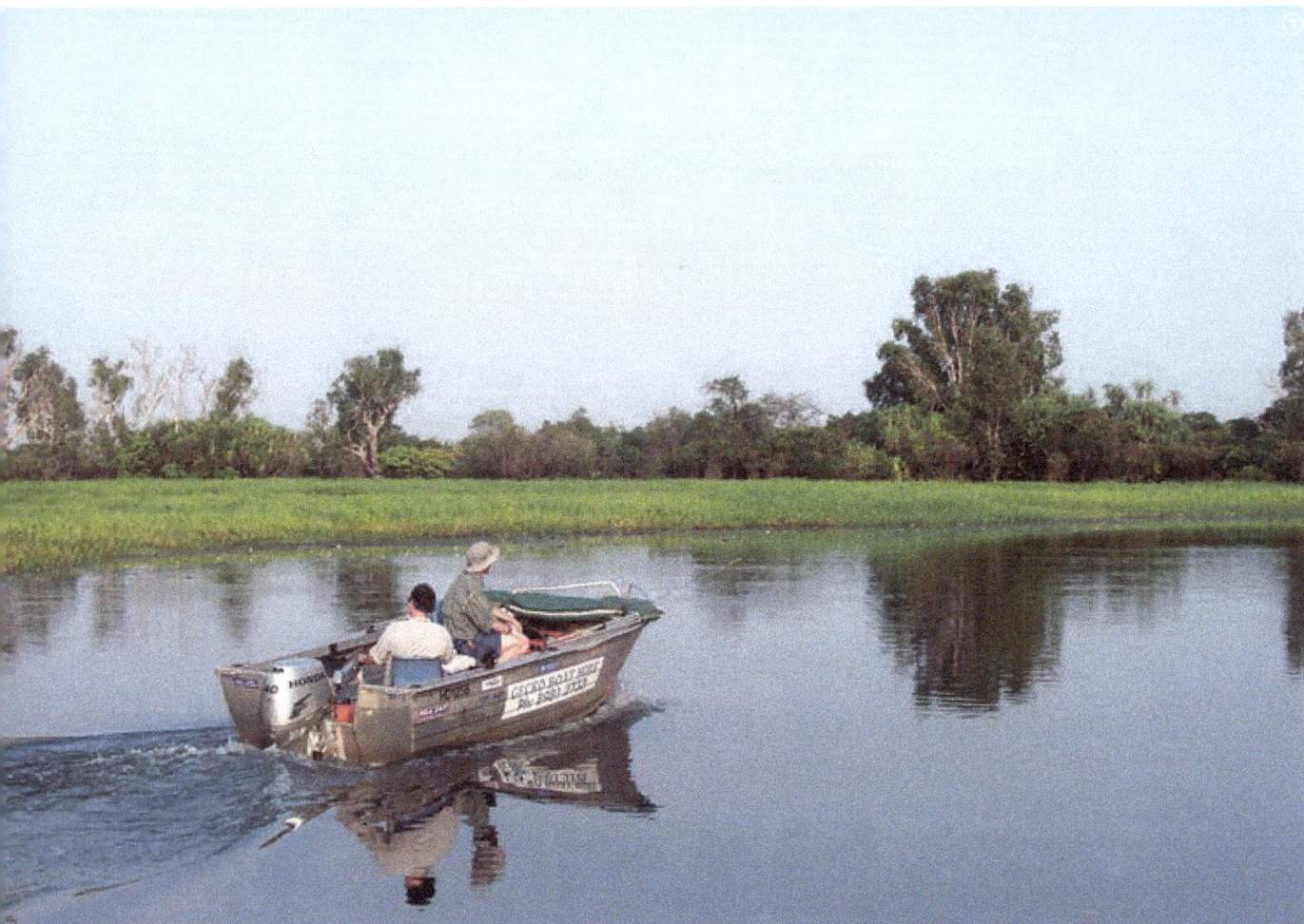

a particularly cost-effective, successful and productive government investment – indeed, one of the few success stories for Indigenous engagement (Gilligan 2006).

Other landholders may also derive an income stream and economic gain from conservation management and the indirect economic benefits that flow from environmental sustainability. Some lands that are currently managed for pastoralism may derive more income from nature tourism or from a biodiversity conservation focus. It is now a reasonable community expectation that all landholders contribute to preventative management of weeds and pests, which in many cases may otherwise pose substantial economic costs to many industries (e.g. the annual cost of weeds to Australia's livestock industries has been estimated at \$315–345 million: Sinden *et al.* 2004). Where such management is beyond a reasonably expected duty of care, funding should be available to landholders for such work through an

expanding range of environmental stewardship options (e.g. Australian government 2007).

Payments to landholders and management agencies for looking after the environments of Northern Australia may be seen by some as a parasitical burden on the economy of the nation. However, as well as the environmental benefits, such payments should be considered in relation to the social benefits they bring to deprived and impoverished communities that characterise much of the North. Additional considerations are the maintenance of a natural asset of tangible value for international tourism and the landscape health on which most resource-dependent industries depend. It can also be justified as an offset for the environmental degradation that has been a consequence of the intensive economic development of temperate Australia.

Increasingly, well-managed and extensive areas of natural vegetation may be expected to provide economic and other social benefits through the

TROPICAL SAVANNAS, FIRE & GREENHOUSE GASES

The tropical savannas of Northern Australia are subject to frequent, extensive fire. These fires produce substantial emissions of greenhouse gases (GHGs), such as carbon dioxide, carbon monoxide, methane and various oxides of nitrogen. Nationally, this savanna burning contributes about 3% of Australia's national GHG emissions. Regionally, these emissions may be very significant – for example, they represent about 50% of the Northern Territory's countable emissions.

Pivotal to understanding how different fires have different effects in the landscape is the 'fire regime' concept. This concept recognises that fires are recurrent, and that the season, intensity and frequency of fire will vary from place to place. Fire regimes are shaped by variations in climate and topography. They are also influenced by the growth rates and flammability of plants. And of course they can be modified by humans.

Emissions of greenhouse gases will vary with fire regime. In Northern Australia, fires in the late Dry season may be extensive and intense, resulting in widespread consumption of dead and living biomass, including litter, logs, grass, shrubs and trees. Intense fires can kill woody plants, particularly trees which store much carbon. All other things being equal, such late Dry season fires result in relatively high greenhouse gas emissions. In comparison, fires in the early Dry season are relatively low in intensity, consume less fuel, and are generally smaller. Consequently emissions of greenhouse gases are lower from these fires than late Dry season fires.

Plants take up carbon dioxide from the atmosphere by photosynthesis, and store it in living tissues (e.g. tree trunks and roots) or in the soil. Carbon dioxide is returned to the atmosphere by respiration (in the plants themselves, and the soil), by animals, both vertebrates and invertebrates (notably by termites), and by fire. The balance between uptake and loss is sometimes referred to as the carbon 'sequestration' potential – if more carbon is taken up than lost over time, carbon is said to be sequestered.

Under the current Australian Greenhouse Office accounting procedures for the national inventory of greenhouse gases, savanna fires are assumed to be carbon neutral. That is, it is assumed that carbon lost by Dry season fires is balanced by carbon uptake the following Wet season. This is not necessarily the case and current research is quantifing the sequestration capacity of savannas, and its sensitivity to variation in fire regime.

Our current estimates (e.g. Chen *et al.* 2003; Williams *et al.* 2004, 2005; Cook *et al.* 2005) suggest that for the *Eucalyptus miniata-tetrodonta* savannas, which are very extensive in the higher rainfall areas of Northern Australia, more carbon is sequestered than lost back to the atmosphere, with this amount increasing when fires are less frequent, intense or extensive. That is, these savannas are carbon sinks. Importantly, the sequestration capacity can be increased if fire frequency is decreased. However, if the savannas are subjected to repeated late Dry season fires, they become a net source of carbon.

Knowing that GHG emissions and the sequestration potential of savannas are sensitive to fire regime, and that fire regimes can be manipulated by people, means that the people of Northern Australia have the opportunity to reduce GHG emissions by managing fire better. This can be done, in particular, by using prescribed fire early in the Dry season to decrease the extent and intensity of fire later in the Dry season, which in turn decreases fire frequency.

Managing fire for abatement of greenhouse gases by decreasing fire extent and frequency brings with it several collateral benefits. These include biodiversity benefits, because a number of plant and animal groups in the savannas are likely to benefit from reduced fire extent and frequency.

There are also commercial benefits. The GHG abatement that could be achieved in places such as Arnhem Land, the Gulf and the Kimberley amounts to millions of tonnes of greenhouse gases per year. This compares very favourably to other national GHG abatement projects which are already operating commercially. For example, in 2004, five million tonnes worth of carbon credits were available to use the under the New South Wales Greenhouse Gas Abatement Scheme (NSW NGAS). GHG abatement in the savanna region through improved fire management is also cheaper per unit CO_2 than many carbon capture and storage schemes in the energy sector (~$10 per tonne vs ~$50). Hence, savanna abatement enterprises may become highly sought after by national and international investors.

These abatement scenarios present real economic opportunities for Aboriginal people, who own extensive areas of the savannas, have the capacity to deliver on-ground fire management, and who are indeed currently engaged in abatement activities. This is an obvious win-win outcome.

There are threats to achieving this abatement potential however. One in particular is the undoubted capacity of exotic pasture grasses such as Gamba Grass to establish, spread and thicken in the savannas. These highly productive grasses lead to hotter, more destructive fires than those fuelled by native grasses. This leads to greater emissions of GHGs and, through higher mortality of trees, a reduction in the capacity of savannas to sequester carbon. This is clearly a lose-lose scenario, and provides an additional reason for controlling the spread of exotic grasses.

Dick Williams, CSIRO Sustainable Ecosystems, Darwin and CRC for Tropical Savanna Management

① Setting fire to vegetation while walking through central Arnhemland: an example of natural resource management that has clear benefits to both human and landscape health.
Photo by Fay Johnston

carbon economy. Markets for carbon trading in Australia are currently under development. In Europe, carbon trades typically from $5–30 per tonne per year (Bayon *et al.* 2007). The savanna woodlands of Northern Australia currently operate as a carbon sink, with about one tonne gain of carbon sequestered per hectare per year (Williams *et al.* 2004). With appropriate land management, these carbon stocks can potentially provide a substantial source of income, given the vast extent of intact savanna woodlands in Northern Australia, especially if a monetary value is placed on protecting extant carbon stocks. This carbon resource can be substantially affected by vegetation clearance, fire regimes, and livestock. The box on the previous page describes one case where fire management to reduce greenhouse gas emissions in savanna woodlands provides a substantial and recognised economic benefit, as offset for a large petrochemical emission source. This case may prove to be a forerunner of a more general situation, whereby the influence of a carbon market shifts the relative economic value of different land uses in favour of enhanced conservation management rather than pastoralism or intensive use based on substantial modification of natural landscapes (Ockwell and Lovett 2005). A further consideration will be the substantial greenhouse gas costs associated with methane from cattle production (about 1.7 tonnes of CO_2 equivalent per head per year, www.ago.gov.au).

STORIES FROM THE LAND

This section provides stories from people making a living from the country in sustainable ways.

The Healthy Country Healthy People project

The future of Australia's landscapes and biodiversity cannot be considered without specific consideration of Indigenous lands and people. Internationally significant landscapes, most within inalienable Aboriginal communal titles, amount to an area the size of NSW in the Northern Territory alone. While Northern Australia has unparalleled opportunities to avoid treading the same path of environmental ruination as southern Australia, without active management, wildfires, weeds and feral animals will degrade these lands.

Indeed, a core problem in managing Northern Australia is the recent depopulation of Indigenous people from their lands that has almost uniformly resulted in unemployment, poverty, ill health and social disruption. Will there be a continuing decline in biodiversity and Indigenous wellbeing in the North mirroring the 19th and 20th Century trends of southern Australia? Or can Indigenous participation in land management activities result in nationally significant benefits to both Northern Australian landscapes and Indigenous health?

The Healthy Country Healthy People project sought Indigenous views about these issues in Northern Australia. In addition it conducted empirical ecological research comparing indices of landscape health under contrasting land management regimes (Indigenous, non-Indigenous and mixed), and epidemiological research evaluating how the health and wellbeing of individuals is affected by engaging in natural and cultural resource management in either traditional or contemporary ways.

The project has provided evidence supporting Indigenous assertions that the wellbeing of people and country are fundamentally connected. Being on country not only provides ongoing and essential management of lands, it provides opportunities for increased physical activity, improved diet, mental wellbeing and fosters cultural and spiritual identity (Franklin *et al.* 2007, Johnston *et al.* 2007). Moreover, higher levels of engagement with natural and cultural resource management were found to be significantly associated with lower risks of diabetes and heart disease, conditions that collectively account for 40% of excess Indigenous mortality (Garnett and Sithole 2007). Similarly a recent review of the Indigenous Protected Areas Program, which supports Indigenous land management for achieving national conservation goals, found that the majority of participating communities reported a range of positive changes including reduced substance abuse, increased school attendance and other indicators of social wellbeing (Gilligan 2006).

Dr Fay Johnston
GP and Medical Health Physician,
Menzies Research Institute

Kaanju Ngaachi Wenlock and Pascoe Rivers, Cape York Peninsula

Homelands and economic development aspirations

I am a Traditional Owner for Kaanju Ngaachi, which encompasses some 840,000 hectares of country centred on the Wenlock and Pascoe Rivers in Central Cape York Peninsula, Northern Australia. We have re-established a permanent community on our Ngaachi at Chuulangun, and have worked hard to ensure the reoccupation of our homelands is sustainable and consistent with Kaanju land management principles. Our ancient governance and cosmology underlie all aspects of Kaanju relationships with homelands and also determine contemporary management of country. Cultural, environmental, economic and social factors cannot be separated as they are all integrated into sustainable land management.

This philosophy is reflected in the comprehensive Management Plan we have developed for our Ngaachi. The Plan sets out the management regime for Kaanju Ngaachi, describing current and future plans for homelands and economic development, including the establishment of an Indigenous Protected Area. An important aspect of our vision is the development of homelands-based economic enterprises that enhance sustainable land management and provide support for our growing homelands community into the future. We are also working with non-Indigenous landowners in the region to enhance sustainable land management and industry to improve livelihoods on country. The establishment of campgrounds and tourism-based activities, the investigation of micro-enterprise based on the sustainable use of Indigenous plant medicines, and a native plant nursery are a few of the many enterprises we are developing on Ngaachi.

Our vision is to conserve, protect and enhance the values of our Ngaachi by way of sustainable land management based on Indigenous governance and cosmology and, at the same time, to benefit economically from our Ngaachi. Integral to the achievement of our vision is recognition and investment from the government and non-government sectors of proper Indigenous governance, land management and the re-establishment of Indigenous people on their particular Homelands.

David Claudie
Kaanju Traditional Owner
Chairman, Chuulangun Aboriginal Corporation,
Cape York Peninsula
www.kaanjungaachi.com.au

2 Fencing of a significant Kaanju cultural site. *Photo by Chuulangun Aboriginal Corporation*

3 Traditional Owners collecting seeds for revegetation project. *Photo by Chuulangun Aboriginal Corporation*

4 David Claudie on the upper Wenlock River. *Photo by Chuulangun Aboriginal Corporation*

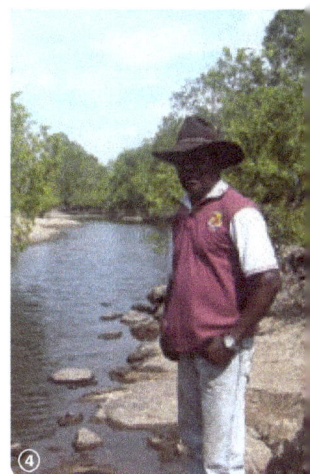

Bininj dja Balanda – working together

Early burning ngarri-wurlhke ka-wurlhme kondah this Arnhem Land. Ngarri-wurlhke bu yekke, bangkerreng ka-yakmen.

Kunwern ngarrih-nani bu TV every city runguhruyi bedberre. Eli arri-wurlhke man-wurrk minj ngarri-djalbawon man-djewk dja man-djewk wardi ka-wurlhme.

Djarre ka-birlire wardi ka-bolkrung, dja Bininj kabirri-rung.

Kamak nganekke ngarri-djare yiman ka-yime ngayi nga-djare. Yawurrinj ba kabirri-durrkmirri kabarri-wokihme kamakNgarrbenbukkan yawurrinj ba bu kabirri-bengkan like ngad maitbi ngarri-danjbik ngarri-dowen.

Wanjh yawurrinj kabirri-bolknahnan kun-red. Karri-djarrkdurrkmirri Balanda dorrengh, ba Balanda kun-wok bedberre dja ngad ngarri-bulerri kun-wok ngarri-wokdi.

Dja Balanda ngandi-bidyikarrme nawu ka-bengkan kun-wale dja kun-wok ngadberre, birri-wern Balanda minj kabirri-bengkan KunwinjkuBu ngayi nga-djare wanjh kunred Bininj ngarri-marnbun, dja Balanda, Balanda law or Bininj law, ngarduk law.

Here in Arnhem Land, we are burning the country early in the year before the hot season. Fires are lit early here.

A lot of times we have seen on the TV, homes in the southern cities have been destroyed by fire. That's why it's important to burn early (after the Wet season) and not to let fuel build up year after year and that is what we do. If this is not done, the fires will travel all over the country for long distances and people will be injured.

This (fire management project) is a good thing and is something we have all wanted. I have really wanted to see this. Proper jobs! Our young people are finding work as rangers. This is a good thing!

We elders need to teach these young people, because when we have died then it's up to young people to look after the country.

And we need to work together with non-Aboriginal people but using both languages. And working with people who are bi-cultural because many non-Aboriginal people do not know our language.

I want it that way. I wanted a place where both Aboriginal and non-Aboriginal people are together – a combination of Aboriginal and non-Aboriginal laws/culture.

(Lofty) Bardayal Nadjamerrek AO

Extract from Catalyst program interview, May 2006, Kabulwarnamyo. Transcription of, and translation from, Kundedjnjenghmi dialect of the Bininj Kunwok language, Dr Murray Garde.

Bringing back the managers

Today a line of longitude and a massive disparity in land management funding separates the stone country of Kakadu National Park from the rest of the Arnhem Land Plateau. The deeply dissected sandstone plateau, with spectacular gorges, jungles, ancient *Allosyncarpia* patches, heathlands and upland forests represents the biodiversity crown jewels of the Northern Territory and is a globally recognised centre of plant diversity.

Despite supporting this extraordinary biological richness, the Arnhem Plateau's soils, climate and ruggedness did not draw the pastoral settlement which saw one group of land managers and land management objectives replaced by others across much of the northern savanna. Instead, the Aboriginal people of the plateau were steadily drawn off their lands from the late nineteenth century until, by the mid-1960s, only a few families persisted on country. Tobacco, medical treatments, flour, sugar and a variety of trade goods drew people. Some found work on pastoral properties, some were encouraged to settle in government settlements and although knowledge of country, both physical and spiritual, persists amongst the plateau diaspora, physical connection and customary land management fell away quickly for many clans.

But now a small group of elders who were born on the plateau, and who represent a precious link with an ancient tradition, are trying to bring back effective Aboriginal management to a largely empty land. A partnership of Indigenous knowledge and science, Aboriginal landowners, government and private enterprise, is underpinning a movement towards a vision of 'healthy country and healthy people'.

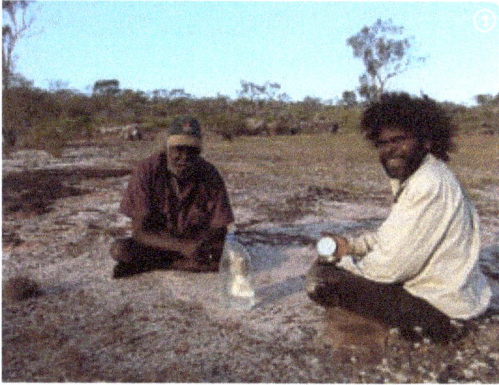

Fire has been a focus for the partnership as scientists and Indigenous experts acknowledge the change to destructive and unsustainable fire regimes which accompanied depopulation. From 1997 to 2006 the NHT invested about $1 million in rebuilding Indigenous capacity to deliver more benign and beneficial fire on the Arnhem Plateau. The Aboriginal organisations of Jawoyn Association, Bawinanga Corporation, Demed Association and the Northern Land Council provided about the same investment and the Northern Territory government's Bushfires NT added critically important science, mapping and remote-sensing.

As the partners developed strategies to reduce catastrophic late Dry season wildfire and replace it with more early Dry season burning, reducing fuel by patch burning and creating long burned firebreaks, the focus shifted to the larger, longer term need – funding labour, operations and capital needs. Through the NT government, the West Arnhem Land Fire Abatement (WALFA) partners have now successfully negotiated agreements providing more than $1 million per year for the next 17 years to fund Indigenous on-ground management. This will be accompanied by the scientific monitoring and accounting to document achievement of a targeted abatement of 100,000 metric tonnes CO_2 equivalent annual emissions from wildfires on the plateau, measured against a ten-year average. The funding is not *from* government but *through* government, as part of emissions offset arrangements with the operators of Darwin's large liquefied natural gas plant, Darwin Liquefied Natural Gas (DNLG). The amount of funding linked to the planned life of the gas plant represents a landmark arrangement between private enterprise and Aboriginal people caring for country on Aboriginal land.

The Indigenous partners have established a permanent land management base at

Kabulwarnamyo on the high plateau to undertake fire and other conservation work. Leaders like Bardayal are bringing back new generations of the landowner diaspora to learn about their country and how to look after it. Endangered Indigenous knowledge is being conserved for sustainable use by this new generation of land managers, who draw on both Indigenous and scientific knowledge systems. Jobs are being created where none have been possible before. Ground work is being laid for other Indigenous landowners

1 Bardayal Nadjamerrek briefs Manwurrk ranger, Len Naborlhborl on where to drop incendiary capsules from a helicopter during the 2007 early Dry season WALFA program. *Photo by Peter Cooke*

2 Bardayal Nadjamerrek and Peter Cooke discuss management issues on the West Arnhem Land fire abatement program. *Photo courtesy of Peter Cooke*

to develop more jobs through tourism, when they feel they are in control of land and cultural management issues. By the end of 2007, landowners plan to declare an area of about 9000 km² of the plateau as an Indigenous protected area, providing management to IUCN specified standards. With extraordinarily high natural and cultural values, and Indigenous management again on the rise, the stage is set for consideration of seeking World Heritage Continuing Cultural Landscape recognition, as landowner confidence and capacity grows.

Peter Cooke

Managing fire in the Kimberley

My name is William Maher. My family came to Yeeda station in 1964 and since then I have worked in the pastoral industry in the Kimberley and the Pilbara regions, especially in the higher rainfall area of the north Kimberley. I have seen big changes in fire management through this period.

The biggest change I think has come with a change in economy. By this I mean that people used to travel over the country a lot more than they do now. The season was started by a team going out to clear around yards and do repair work to them, gather firewood for the branding fire, and pull trees off roads if you had no grader. All the time you would be burning as you went. These fires would be quite small and fairly cool. The easterly winds were not going at this time.

Then would come the mustering team covering the country quite extensively. You would burn as you went, and the fires were bigger and hotter as the easterly wind was going well. The ground still had a lot of moisture in it and encouraged good regrowth. These fires would be stopped by the patches of early burns in some cases and in others by creeks that would still have water in them and by sparse areas of grass that had been burnt the year before.

The next thing in the fire department was thunderstorms. These would come in the early Wet season and could be extensive if you had a series of dry storms. Others would cover much smaller areas if the storms had good rain in them.

The difference these days is that we now spend a lot less time burning the country early in the year. The moving around is done by aircraft when managers wish to look over the land. Time now is spent doing other things that are more productive to the way things are set up. Wages are so high and labour is so hard to find that some things are now excluded from the day-to-day running of the station.

So in this cyber age we see a lot of fire on our computer screens whereas 30 years ago we only saw what came over the horizon. The satellite technology can be a great tool or it can also be very frightening to the uneducated. I think that the fire permit system has created more fire than it has prevented. People are so worried about prosecution that they will not do preventative burns where otherwise they might have. I think these changes with fire are causing all sorts of problems with wildlife.

I have been involved with fire management in one way or another for a lot of my life. I am a licensed chopper pilot, and some of my time now is spent with contract aerial incendiary work. I believe that smaller fires spread throughout the year, creating varied fire histories, are preferable to one large burn, and I hope I help achieve this with the chopper service.

There are very few people left in the area who have knowledge and a history of fire usage, few who can educate youngsters and newcomers.

Butch Maher

Indigenous Protected Area program – Groote Eylandt

The Indigenous Protected Area or IPA program is an initiative of the federal government (www.environment.gov.au/indigenous/ipa/index.html). It fosters the long-term cooperative management of premium Aboriginal-owned land for the preservation of cultural and ecological values. The IPA program is nested within the broader National Reserve System, which aims to establish comprehensive, adequate and representative samples of Australian bioregions. A prerequisite for an IPA is a plan identifying agreed objectives, actions, guidelines and standards for land management. In general, these are closely aligned with other state, territory or federally managed conservation areas, meaning IPAs are managed to the same standards as a National Park.

Benefits of an IPA are broad. Fundamental is the commitment to the long-term protection, sustainability and integrity of environmental and cultural values. It recognises conservation as a legitimate and worthwhile land use and, importantly, provides financial benefit. Development of an IPA and subsequent management actions encourage and facilitate extensive Indigenous participation and engagement with government and other NRM agencies. This promotes a recognition, understanding and appreciation for Indigenous land management practices. IPAs provide a mechanism whereby traditional ecological and cultural knowledge can be synthesised with contemporary scientific approaches to maximise outcomes. Further, exposure to western land management practices and science provides opportunity for Indigenous skills development and vice versa.

Complementary to this are notable social benefits. Aboriginal people often live within a community struggling with elements of social dysfunction. Poor health, low standards of education, unemployment, substance abuse, higher rates of crime and incarceration are sadly far too common. Among this background, an IPA program offers real jobs, money, opportunity and hope. Aboriginal people are closely affiliated with country, and often have a cultural obligation as guardians of their land. Working as a ranger, or being involved in an IPA provides one opportunity to facilitate this requirement. Feelings of pride, satisfaction, wellbeing and happiness can be all associated with environmental work. For individuals, IPAs offer opportunities for training, chances to exchange ideas and experiences with other rangers, and ultimately self improvement. In time, a functioning IPA ranger group engages with schools, educating children about the value of conservation.

Simon Hartley

A case study: the Anindilyakwa (Groote Eylandt) IPA

The Anindilyakwa archipelago includes approximately 40 islands, and is located in the Gulf of Carpentaria about 630 km east of Darwin. All land is owned by Anindilyakwa speaking (the local language) clans. The main island, Groote Eylandt, is Australia's third largest island and covers an area of about 2687 km^2. In the 2001 census the population was estimated to be 2419, of which approximately 1500 were Indigenous.

The Anindilyakwa archipelago is in an enviable position in the context of Australia's, and indeed the world's, natural environment. The area is relatively biodiverse with more than 40 mammals, 70 reptiles, 15 amphibians and more than 200 bird species. Included in this group are some threatened species such as the Northern Quoll, Brush-tailed Rabbit-rat, Northern Hopping-mouse and several species of marine turtle. Whilst there is an impressive list of endemic and threatened wildlife, probably of more significance is the absence of feral animals from the island. This is probably the largest area in Australia without exotic grazing mammals, and there are no pigs, buffalo, foxes, horses, cattle, goats and cane toads.

Complementing the archipelago's wildlife is a plant community which hasn't been farmed, grazed, forested or cleared in a major capacity and, whilst there is a small amount of mining, impacts are localised and there is a rehabilitation program. Surrounding the island is a marine

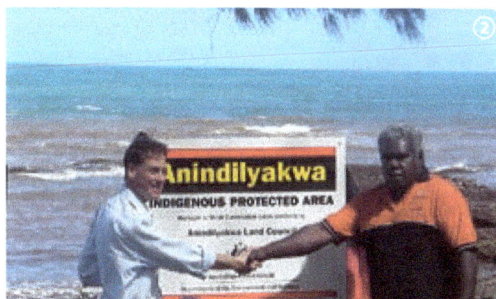

environment with stunning scenery, fabulous reef systems and rich Indigenous, commercial and recreational fisheries. Accompanying all of this is an Indigenous population that speaks traditional language, harvests bush tucker and medicines, and still practises long-established customs and rituals. Having all these qualities in a single place is rare in the context of contemporary Australia and truly makes Groote a unique and highly valuable location in terms of environmental values, and well-qualified as an IPA.

The Groote archipelago was designated an IPA in mid-2006, following implementation of a plan of management by the Anindilyakwa Land Council (ALC) from 2005. Six Indigenous rangers, a coordinator and a project officer are employed to execute the plan of management. Their duties are broad, and in the last 12 months have included intensive training, construction of a board and chain road, surveys for threatened species and biodiversity, trapping of pest animals, monitoring of turtle populations, visitor management, patrols for illegal fishing vessels, crocodile removals, talks to school children, and the collection, processing and recording of ghost nets and marine debris. This work is underpinned by IPA and ALC funding, but could not be sustained without external funding, which requires constant applications, presentations and reporting.

The rangers initially took to the program with vigour and enthusiasm. The first 12 months were productive in terms of on-ground activity and skill development. For individuals there were some pleasing outcomes. Qualifications such as a driver's licence, coxswain's ticket, chainsaw tickets and first-aid courses were lifetime firsts. For some, a commitment to a work ethic and sense of achievement was evident. They had pride in their performance, developed self-esteem and demonstrated through more than 20 talks to the community about their work. As one of the rangers said, 'This work provides me with a reason for being'.

Support offered by the employer, project officer and coordinator assists with an understanding of mainstream Australian lifestyle skills such as banking. The collection and recording of data demonstrated the value of literacy and numeracy and fostered a desire for knowledge. The attendance of rangers' kids at school increased and anti-social behaviour decreased.

Some benefits, especially the social ones, could have been delivered by any worthwhile work program; however, the reality is that looking after country is culturally important to Aboriginal people, and therefore makes the IPA program more likely to deliver. The Anindilyakwa IPA is relatively new but, with continued support and hard work, the program has the ability to deliver significant environmental, cultural and social benefits.

Simon Hartley

Mornington Wildlife Sanctuary, central Kimberley

Mornington Wildlife Sanctuary sits squarely in the middle of the Kimberley, and is blessed with a spectacular landscape. The heavily rippled sandstone layers of the King Leopold Ranges arc through the southern part of the property. North of this, massive mesas and smaller volcanic hills stand majestically over the sweeping savannas. Monsoonal rains are collected in a network of rivers and creeks that coalesce in the heart of Mornington to form the Fitzroy River, which then squeezes through kilometres of magnificent sandstone gorges to reach the expansive Lower Fitzroy floodplain country.

Besides a dramatic landscape, Mornington has outstanding natural values, including healthy populations of several threatened animals, like Northern Quolls and the dazzling Gouldian Finch, as well as ecosystems that are not protected in National Parks. Before the Australian Wildlife Conservancy (AWC) bought it in 2001, Mornington was managed as a pastoral lease.

AWC is an independent, non-profit organisation dedicated to the conservation of Australia's threatened wildlife and ecosystems. It owns 15 properties nationally, covering more than 1,100,000 ha. Five of these (covering 670,000 ha) are in Northern Australia. AWC invests an enormous effort (sometimes with the help of partner organisations) into on-ground management, conservation-related research, and educational visitor programs, all of which are funded by tax-deductible donations, augmented with grants for specific projects.

Mornington is the largest AWC sanctuary, at over 350,000 ha. Eight to ten permanent staff live onsite to implement the fire management, destocking, feral animal and weed control, as well as carrying out a suite of biological

monitoring and research programs designed to improve the land management. The Traditional Owners of the area, who live on Mornington at Tirralintji, are actively involved in these land management programs.

During the Dry season, a small visitor facility – a combination of a luxury tented camp and a campground – is run by an additional ten seasonal staff. As well as giving the 3500 annual visitors an opportunity to experience the landscape and wildlife of Mornington, the Wilderness Camp provides a program of guided and self-guided walks, evening slideshows, and a small interpretation centre that showcases the AWC mission and the conservation issues facing Northern Australia.

Mornington provides a nexus for ecologists, land managers, Traditional Owners, travellers and philanthropic supporters to express their care for the environment, and is an example of a sustainable land use alternative in Northern Australia.

Sarah Legge
Ecologist, Australian Wildlife Conservancy

❶ Kim Hands *(second from left)* guiding Mel Berris *(left)*, Emma Flaxman and Annette Cook on Mornington Station. *Photo by Alex Dudley*

❷ Brush-tailed Possum captured in monitoring program, Top End. *Photo by Brooke Rankmore*

① Brahman Bull.
Photo by Barry Traill

Lessons from history – Valuable for the future

After nearly 100 years and four generations of one family on one property, valuable lessons have been learnt regarding successful management techniques for profitable production and environmental sustainability. Our property, Trafalgar covers 33,000 hectares and is situated in the semi-arid tropics near Charters Towers in North Queensland. We specialise in beef production, but are passionate about natural resource management, as we believe it underpins not only successful beef enterprises, but also a healthy environment generally.

When my grandfather bought the property in 1913, it was relatively undeveloped with few waters and very little fencing. The cattle herd was all British breeds and had little tolerance of drought conditions or parasites such as the cattle tick. As a result, mortality rates were high, so overgrazing rarely occurred and pasture integrity was maintained. During the 1960s, the tropical breed, Brahman, was introduced along with new technologies like supplementary feeding and some exotic pasture species. Brahmans have a much higher tolerance for drought and parasite resistance so the mortality rates experienced with British breeds were greatly reduced.

During the beef price slump of the 1970s high cattle numbers were maintained largely due to a run of good wet years in conjunction with the feeding technologies now widespread in the industry. It was in the mid-1980s that producers, scientists and other interested groups realised the ramifications of the then current management practices. Pasture degradation, in the form of loss of perennial species, decreasing basal area, exotic weed intrusion and erosion due to bare areas, was highly evident. Because of the low pasture yields, preparing for drought was not an option, so it also became a high-risk enterprise. On Trafalgar, we reduced our stocking rate by 60%. Then by spelling at least 20% of the property every Wet season, we were able to restore native pasture species to greater than 80% within a few years.

These lessons have now led us to our current management regime, where spelling 20% of the property annually, strategic use of small areas of exotic pasture, conservative stocking rates and intensive herd management have increased our productivity (i.e. higher calving rates, earlier and heavier turn-off weights, better meat quality) and therefore profit. Monitoring sites on the property also confirmed the improvement in pasture quality, soil health and water quality. We also have an annual control program for exotic weeds. Current research in natural resource management also confirms these strategies lead to improved biodiversity and ecosystem health.

Because of the high amount of biomass available in the pastures, there is very little risk attached to the enterprise in the event of a failed Wet season. In other words, there are always options whether it be drought mitigation, the use of fire for weed control or pasture regeneration or taking advantage of low cattle prices to buy in stock to fatten.

As farming land in Australia becomes more expensive, more pressure is going to be applied to existing land, particularly marginal grazing areas. Planning at a national, regional and property level is imperative, if we are to combine best practice and research to achieve sustainable outcomes for the Australian landscape and its inhabitants.

Roger Landsberg
Trafalgar Station Charters Towers

ECONOMIC ACTIVITIES

There are diverse existing interests in the future of Northern Australia. In Table 6.1 we list a range of land use activities and classify them in terms of their effect on natural values and ecological processes as either *conservation, compatible, potentially compatible* or *incompatible*. This classification draws upon documented experiences with environmental problems from unsustainable land use activities both in the North and in southern Australia. While this assessment is qualitative, it identifies the different potentials for sustainability, and for serious and irreversible environmental harm. As shown in this table, compatible activities may be completely or largely separate from country (e.g. information technology), or may use country but with minimal impact (e.g. nature-based tourism).

Accordingly, we believe that certain proposed developments, such as broad-scale clearing,

TABLE 6.1 CLASSIFICATION OF COMPATIBLE & INCOMPATIBLE LAND USES

Classification of economic activity	Description	Examples
Conservation	Activities that directly and actively help to maintain ecological processes and natural values.	• Some government services (border control and quarantine) • Conservation management of country • National Parks, Indigenous Protected Areas, off-reserve management • Feral animal control and harvesting
Compatible	Activities that rarely if ever degrade, but may simply be neutral to the environment.	• Most government services (defence, health, education, provision of infrastructure) • Visual and creative arts, including the Indigenous arts and crafts sector • Nature and culture-based tourism • Information technology services • Biotechnology (e.g. bioprospecting); for example, identifying new medicines using traditional knowledge
Potentially compatible	Activities that can be compatible with maintaining natural values and processes if done with care and in particular ways.	• Pastoralism dependent on native pasture and operating within carrying capacities • Mining operations that have minimal water requirements and small ecological footprints • Low input aquaculture; for example, with natural feeding stock (e.g. shellfish, sponges) • Harvesting of native plants and animals from the wild • Fishing (commercial, recreational and Indigenous) • New residential and tourist developments • Mass tourism • Military training • Trophy hunting of feral or native animals
Incompatible	Activities that are inherently degrading to natural values and processes, for which significant damage can only be reduced, and generally not to a satisfactory level.	• Permanent and large-scale clearing of native vegetation for agriculture • Mining operations with a large ecological footprint (such as strip mining without adequate rehabilitation or protection of hydrological systems) • Large-scale water off-takes, impoundments and irrigation • Extensive plantation development • Extensive aquaculture developments with high input (e.g. fish fed other fish) • Genetically-modified crops • Pastoralism using invasive introduced grasses

which are *incompatible* should be rejected. This recommendation will be resisted by some, including those with vested interests. However, the exclusion of incompatible land use activities will remove many of the large-scale damaging threats to natural values and ecological processes, thereby providing the basis for ecologically sustainable development.

We have attempted to capture the possible range of environmental compatibility within different land uses in Figure 6.1. For many activities, the manner in which they are carried out will determine their category. Conservation management in almost all areas of Northern Australia now needs people on the ground, delivering management services, in terms of weed, pest and fire management. For land use activities potentially compatible with the sustaining environmental values, the challenge is to progressively shift into the positive side of the ledger. For society as a whole, the question is how best to support industries to do that.

❶ King George Falls cascades 80 metres off the Kimberley Plateau into the saltwater reaches of the King George River on the North Kimberley Coast. *Photo by C Hugh Brown*

❷ Cattle mustering, Mornington Station, Kimberley. *Photo by Alex Dudley*

Pastoralism

Pastoralism – largely for beef production – is by far the dominant land use in Northern Australia, covering more than one million square kilometres. The industry can potentially have major impacts beyond boundary fences of properties, through the escape of exotic pasture grasses to lands of other tenure, and through catchment-wide impacts on water flow and quality. Given this, the sustainability of Northern Australia's ecological assets is contingent upon how the pastoral industry is managed.

Pastoralism has a history from the beginning of European settlement of Northern Australia. Unlike industries with more conspicuous, localised impacts, there is far less public appreciation of the environmental costs associated with pastoralism and assessment or explicit regulation of that impact. Nonetheless, all northern jurisdictions are now engaging in policy and legislative processes to review conditions of pastoral leasehold lands. In some cases, these reviews may provide for greater environmental accountability of pastoral landholding. But, against this momentum of higher community expectations for environmental stewardship of pastoral lands, there is an increasing drive for intensification of land use. In part, this drive is a response to rising operational costs that force landholders to try to derive more income from their leases. This may mean more cattle at densities which are much closer to (or above) carrying capacity. Increased stocking rates are achieved through developing infrastructure (such as for smaller paddock sizes), increased use of invasive foreign grasses (that often need tree clearance to establish effectively), and greater use of water.

Increasing the intensity of land use is not generally compatible with protecting biodiversity and maintaining the ecological processes that underpin healthy country. With respect to the pastoral industry in Northern Australia, implementing our guiding principles requires that:
- There is an adequate representation of all environments within lands managed primarily as conservation reserves. Currently, some more productive grassland types are almost exclusively held by the pastoral industry;
- Any substantial intensification on pastoral lands be assessed through an environmental impact process, with due attention given to the value of natural environments and the risks posed by land use change;
- The carrying capacity of pastoral lands be explicitly defined, with inclusion of the full gamut of biodiversity attributes, and never exceeded;
- An adequate level of duty of environmental care should be expected of pastoral landholders;
- Incentive programs should be available to encourage landholders to extend their environmental standard beyond the duty of care benchmark;
- Non-pastoral use (explicitly including management for conservation benefit) should be encouraged, at least in part to increase the long-term security of Northern Australian economies through enterprise diversification;
- Clear disincentives should be applied to landholders who fail to exhibit reasonable duty of environmental care; and
- 'Best Management Practices' (BMP) should be developed for the industry. BMPs can be used to establish explicit and attainable environmental goals and timetables that are regularly updated to reflect changing conditions and new knowledge.

Indigenous lands

Aboriginal and Islander lands comprise nearly 20% of Northern Australia, and this will increase in the future as further pastoral lands are purchased by Indigenous agencies, and an increasing proportion of conservation lands are returned to Aboriginal ownership or joint management. Aboriginal lands make a major contribution to the natural value of Northern Australia. A complex set of factors affect the management of these lands. The most dominant of these factors is the socio-economic disadvantage characteristic of most Aboriginal communities. The local economies are usually limited, with few employment opportunities, inadequate resources available for management, and health and education standards are poor. There are strong cultural drivers for Aboriginal people to live on remote country and maintain traditional management and culture. Conversely, there are economic and social drivers for people to join the mainstream, live in more centralised communities and open their lands to more intensive economic activities (such as for plantations, mining and pastoral enterprises). Relative to the North's non-Indigenous population, Indigenous birth-rates are high and emigration rates low. There is an inexorable demographic trend that Aboriginal people

FIGURE 6.1 **ENVIRONMENTAL CAPABILITY OF DIFFERENT LAND USES**

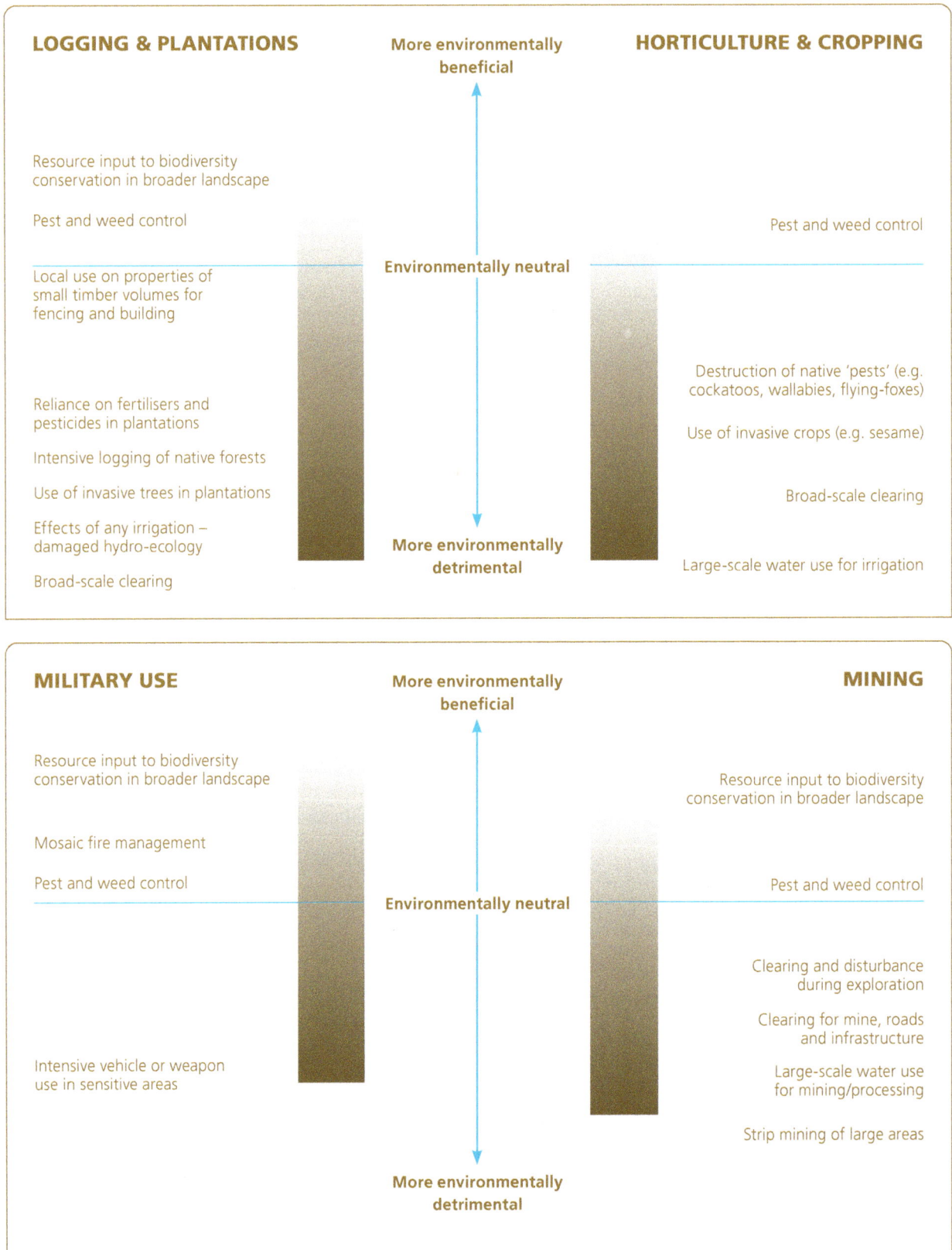

LOGGING & PLANTATIONS

More environmentally beneficial

HORTICULTURE & CROPPING

Resource input to biodiversity conservation in broader landscape

Pest and weed control

Pest and weed control

Environmentally neutral

Local use on properties of small timber volumes for fencing and building

Destruction of native 'pests' (e.g. cockatoos, wallabies, flying-foxes)

Use of invasive crops (e.g. sesame)

Reliance on fertilisers and pesticides in plantations

Intensive logging of native forests

Use of invasive trees in plantations

Broad-scale clearing

Effects of any irrigation – damaged hydro-ecology

More environmentally detrimental

Large-scale water use for irrigation

Broad-scale clearing

MILITARY USE

More environmentally beneficial

MINING

Resource input to biodiversity conservation in broader landscape

Resource input to biodiversity conservation in broader landscape

Mosaic fire management

Pest and weed control

Pest and weed control

Environmentally neutral

Clearing and disturbance during exploration

Clearing for mine, roads and infrastructure

Intensive vehicle or weapon use in sensitive areas

Large-scale water use for mining/processing

Strip mining of large areas

More environmentally detrimental

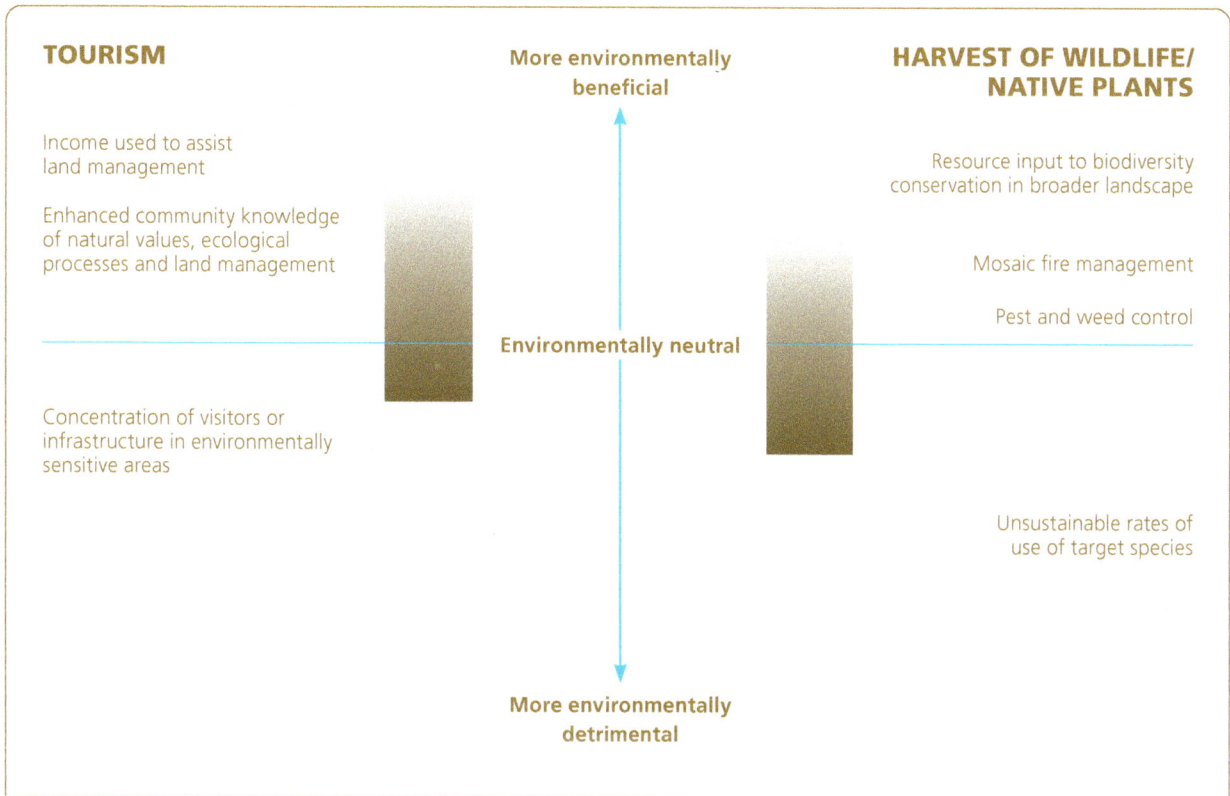

TOURISM

Income used to assist
land management

Enhanced community knowledge
of natural values, ecological
processes and land management

Concentration of visitors or
infrastructure in environmentally
sensitive areas

**More environmentally
beneficial**

Environmentally neutral

**More environmentally
detrimental**

**HARVEST OF WILDLIFE/
NATIVE PLANTS**

Resource input to biodiversity
conservation in broader landscape

Mosaic fire management

Pest and weed control

Unsustainable rates of
use of target species

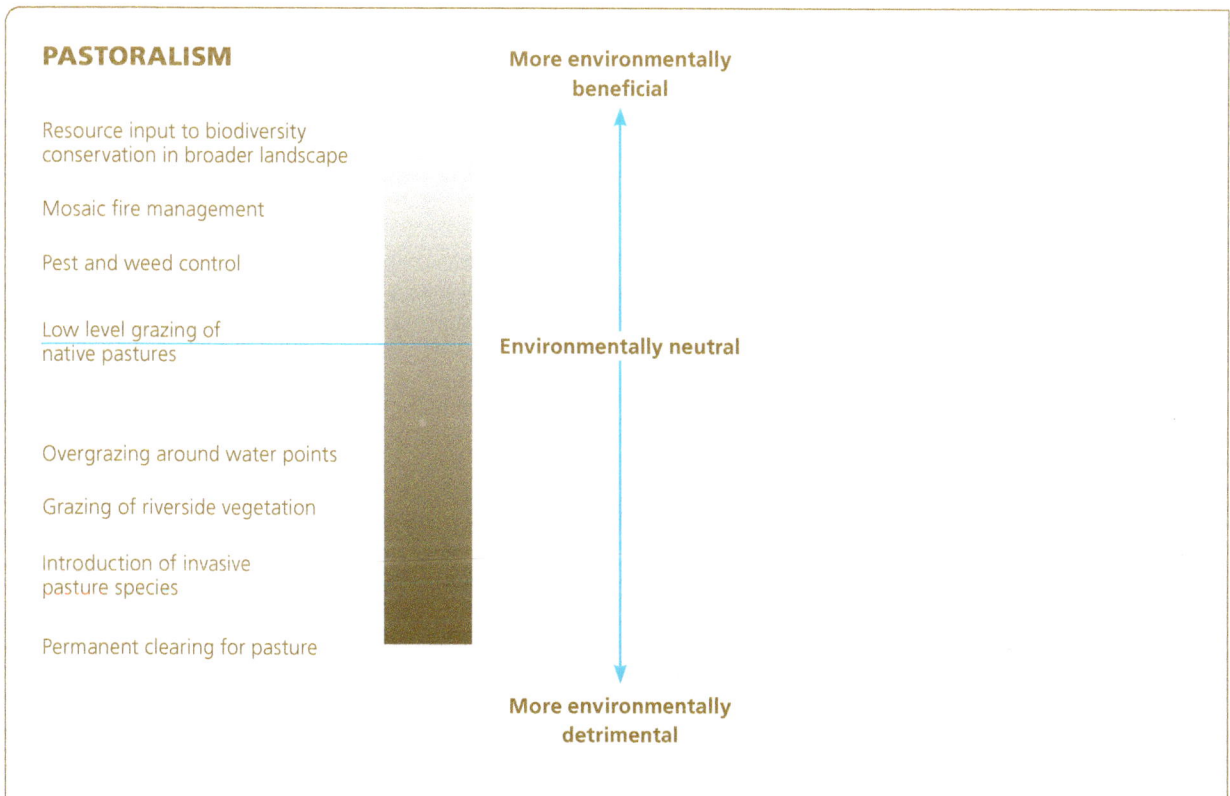

PASTORALISM

Resource input to biodiversity
conservation in broader landscape

Mosaic fire management

Pest and weed control

Low level grazing of
native pastures

Overgrazing around water points

Grazing of riverside vegetation

Introduction of invasive
pasture species

Permanent clearing for pasture

**More environmentally
beneficial**

Environmentally neutral

**More environmentally
detrimental**

will comprise an increasing proportion of the North's population, and will be the demographic group with the most long-term residents.

A sustainable future for Northern Australia is tied to Indigenous people and lands. To best nurture that future, substantial resources must be invested in providing jobs and education for Aboriginal people to manage, live in and sustainably use their lands. Such use may include conservation partnerships (such as joint-management of National Parks and the development of Indigenous Protected Areas), carbon trading, sustainable economic use of resources for the art/craft industry, for bush foods, for safari hunting and other modes of tourism, and harvest of native foods for local consumption. But, inevitably, there will also be demands and opportunities for intensive development. This may be an appropriate and reasonable source of socio-economic benefit so long as (a) it does not compromise the environmental fabric that underlies the sustainable future of Northern Australia; (b) development constraints are not framed by Third World standards (e.g. approval conditions must not be weakened to allow for the local and short-term financial benefit of a development that has poor environmental credentials); and (c) socio-economic returns are not distorted to the current low economic standards of communities (e.g. rents paid to Aboriginal landowners for land use should not be less than those paid for similar provision by landowners elsewhere in Australia).

Conservation lands

National Parks and other conservation lands comprise about 6% of Northern Australia, appreciably less than basic international and national benchmarks (Sattler and Glanznig 2006). The overall standards of management, resourcing and monitoring of performance and biodiversity retention are variable. With respect to protected areas in Northern Australia, implementing our guiding principles requires:

- A substantial expansion of the core reserve system to at least meet national objectives;
- Landscape-wide planning to maintain ecological connections between core reserves and facilitate collaborative management of threats across land tenures, including enhancement of the formal government reserve system, and non-government reserves such as Indigenous Protected Areas and privately owned conservation reserves, conservation covenants and other off-reserve initiatives;
- Substantially enhanced resourcing for the parks network; and
- Explicit management planning to ensure that reserves deliver clearly articulated biodiversity conservation outcomes.

Military lands

Lands managed for military operations and training constitute nearly 1% of Northern Australia. Such lands may provide considerable conservation outcomes (Woinarski and Ash 2002; Woinarski *et al.* 2002), because they often have substantial resources available for land management, no livestock and management is explicitly tied to Commonwealth environmental standards under the *Environment Protection and Biodiversity Conservation Act*. However, they may also be susceptible to local intensive and damaging use (such as areas in which live firing is concentrated), and there may be little scope for public accountability. Conservation benefits of these lands will be maximised when the management planning adequately addresses biodiversity issues, when management of weeds, pests and fire is approached within a broader regional context, and when the sustainable limits of intensive use are not exceeded.

Mining

Mining is the most important non-government economic sector in Northern Australia. Its characteristics, scope and impact vary considerably, depending upon the mine type, duration, extent and location. The most substantial impacts occur with strip-mining (currently for bauxite at Weipa and Gove, and for manganese on Groote Eylandt), which requires extensive vegetation clearance and the risk of substantial impacts to hydrology, soils and local biodiversity. Some individual smaller mines produce a risk of long-term pollution through leaching of toxic materials. In certain regions, while the impacts of individual mines may be limited, a procession of mines may lead to substantial cumulative impacts. While recognising that Northern Australia will continue to have a substantial mining industry, implementing our guiding principles will require acceptance that:

- There are habitats, places and regions in Northern Australia where the environmental values and ecological integrity are so profound that no mining should be countenanced;
- For any mine, there must be an explicit condition of no off-site impact;
- For any mine, there must be an explicitly-defined objective to return at the end of the mine-life to pre-mining natural environmental state, with adequate and assured resourcing to provide for such return and adequate independent auditing to evaluate such return; and
- A proportion of the mining economic benefit be allotted to regional provision of environmental offsets, to ensure that there is no net loss of environmental values due to the mining venture.

Horticulture and plantations

Horticultural production, including plantation forestry, is a steadily escalating land use in Northern Australia, stimulated recently by an increased awareness of the environmental fragility of horticultural production zones in southern Australia. The history of horticulture in Northern Australia is chequered, with some notable instances of failure and environmental degradation (Bauer 1977). Many failures were due to lack of appreciation of the environmental characteristics and constraints in Northern Australia; particularly related to the limitations and challenges of climate (including cyclones and floods), water availability, and very limited areas of fertile soils. Notwithstanding the recurrent spruiking, and more insistent recent interest, the ecological reality is that the horticultural potential of Northern Australia is limited. For those regions where intensive horticulture may have prospects, such development may come

at considerable environmental cost. Hence, to implement our guiding principles in Northern Australia, horticultural development should:

- Not proceed in habitats, places and regions in Northern Australia where the environmental values and ecological integrity are profound;
- Be evaluated within a regional context that first ensures adequate lands and water are set aside for the maintenance of environmental values and functions;
- Have no off-site impacts; and
- Include provision of regional environmental offsets, to ensure that there is no net loss of environmental values due to the horticultural development.

WHERE TO FROM HERE?

We have not attempted here to provide a detailed plan for appropriate development and conservation management across Northern Australia. Such a task is beyond the scope of our study. This is the responsibility of all people across the North, and of all Australians interested in its future. Furthermore, a variety of legal, programmatic, fiscal and policy mechanisms must still be developed to provide the framework on which to base a sustainable future for the North; including voluntary covenants, co-management arrangements, markets for new ecosystem services, and private investment in new green industries.

1 Bauxite mine, Weipa, Cape York Peninsula. *Photo by Kerry Trapnell*

2 Mango farm near Darwin, Top End. *Photo by Barry Traill*

107

INCREMENTAL INDUSTRIAL ACTIVITY ON THE KIMBERLEY COAST

The current industrialisation of the Kimberley coast offers a text book example of how incremental decision making can occur without assessment of the combined impacts.

The Kimberley coast (and its immediate inland areas) is one of the most isolated parts of Northern Australia. It is well-known for its spectacular scenery and its natural values remain highly intact. Off the coast is the Browse Basin, with large and so far untapped natural gas fields. A number of companies are in the highly advanced stages of planning new industrial development along the Kimberley based on the resources of gas, and minerals on the adjacent lands of the Kimberley.

These proposals include: offshore gas production and building of two new large liquefied natural gas plants at coastal sites, two bauxite mines, two ports, an alumina smelter, a zinc mine and smelter, and an iron ore mine on an island.

Many or all of these proposals offer the potential of significant wealth for Australia as a whole. However, each single proposal is likely to have significant local environmental impacts, and local social impacts. In combination, the plans mean that the Kimberley coast potentially faces considerable industrial development along its length. Many community groups have called for a regional planning process to assess the whole basket of development.

Despite the likelihood of the accumulating impacts on the environment, at the time of writing there has been no attempt to carry out a region-wide analysis or consultation on the cumulative, region-wide impacts of these proposals. As has been the pattern with this approach elsewhere the combined incremental changes that may occur could have a severe impact on the nature of the Kimberley. Piecemeal assessments are unlikely to produce ideal conservation and development outcomes.

Barry Traill

Our focus here has been to provide information on the natural values and ecological processes of the North so that they can be understood, appreciated, and serve to frame guiding principles for delivering a sustainable future. However, we can usefully make some further comments on the requirements for sustainable development pathways. As noted, current conservation and development planning across the North usually follows an approach of incrementally assessing development and conservation.

We have suggested that this approach is a pathway to incremental degradation, and to a development portfolio that is environmentally unsustainable. An example demonstrating the problems of this approach is detailed in the side box on Kimberley Gas developments.

Much more sophisticated regional planning processes are required if we are to build development pathways that work for people and country.

And how do we deal with specific industries? In the text and figures above we explain how some economic activities are incompatible with the long-term protection of the natural values of the North. We have also explained how many can be more or less compatible depending on the way in which they are conducted.

Where currently present, those existing incompatible activities should not be extended. Rather, alternative approaches to economic development should be pursued when possible.

For activities where compatibility with the long-term maintenance of land and water is possible, pursuing pathways to improve sustainability is already in place in many industries, such as in pastoralism. Pursuing environmental Best Management Practices at an industry level provides an approach that has potential to deliver major improvements to performance.

The challenge now and in the coming decades is to maintain the natural values of the North, protect the ecological processes that sustain these values, and repair any environmental damage that has already occurred. A large part of meeting this challenge will involve promoting a shift in the kinds of land use activities that occur in Northern Australia, and how these are carried out. The contributions of Indigenous and pastoral communities will be crucial, as they are the majority of owners and active custodians. If we fail in this challenge, then it is inevitable that the North's natural values will erode over time, leading to the kinds of environmental problems now dominating the South – water security, species extinctions, land degradation and loss of agricultural productivity.

SUMMARY

In summary, our analysis of the nature of Northern Australia has been in three parts.

First, we document that Northern Australia has natural values of great significance. In particular, it has the largest and most intact tropical savanna left on Earth, and the majority of Australia's remaining natural rivers and associated wetlands. It also has nationally important areas of rainforest, mangroves, tropical heathlands and other habitats. It has many species found nowhere else in the world. This natural landscape provides valuable ecological assets on which major northern industries such as tourism, fisheries and pastoralism are based.

Second, we outline and illustrate the importance of ecological processes and connections that link and support nature, environmental health and ecological functioning in Northern Australia. Unlike southern Australia, in the North these processes remain largely intact. Hydro-ecology (water), disturbance (fire), and long-distance movements of wildlife are the key ecological processes at work. Maintaining these processes is a foundation of maintaining healthy ecosystems and the people they support across the North.

Third, we propose a model for shifting development in the North onto a pathway that maintains, supports and protects the natural values and supporting ecological processes of the North:

- A regional planning process that identifies the capacity of regions to absorb human-induced changes to the landscape;
- The establishment of core areas to be managed primarily for conservation;
- Constraints on activities that are directly or indirectly destructive to the natural values and ecological processes of the North;
- The promotion of economic activities that are, or can be made to be, compatible with those values and processes;
- The promotion and coordination of management compatible with conservation across all land tenures;
- Fostering collaborative approaches to conservation and management amongst landholders; and
- The facilitation of a 'conservation economy'; enterprises that yield a net positive gain for the natural environment.

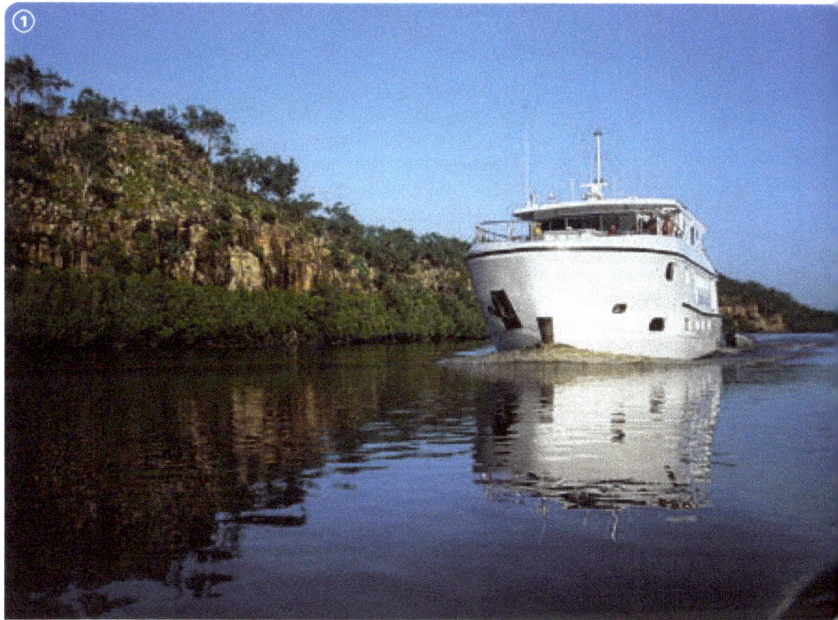

The diagram on the following pages illustrates this model by showing alternative approaches to land use in Northern Australia.

1 The Kimberley Quest II brings travellers to the King George Falls, Kimberley Coast.
Photo by C Hugh Brown

CONCLUSION

Over the course of our history, Australians have grappled with working sustainably on our continent. In many regions we have demonstrably failed. Our generation, and future generations, are now recognising the consequences of unsustainability, and the price of attempting to restore over-used environments. In other regions, such as the North, collectively we now know enough and are wealthy enough, to not follow the same pathways. We can consciously make decisions to look after country, and support the people of Northern Australia in the process.

The natural environment of Northern Australia has undergone extraordinary changes since humans first arrived. Many of these changes have been the result of Indigenous land use activities, framed within traditional cultural obligations to manage country. But every plant and animal species alive today in the North evolved millions of years ago and has persisted through the rise and fall of sea levels, and periodic global warming and cooling. And, as we are only now beginning to appreciate, they have survived for the last 50,000 years in an intimate relationship with their Indigenous custodians. What will be our legacy?

HEALTHY COUNTRY IN THE NORTH

Sandstone plateaus
and escarpments

Healthy springs keep
waterholes full during the Dry

Water from
springs maintain
river flows in
Dry season

Indigenous rangers on
outstation manage country

Fruit-bats move over
savanna, seeking nectar
and pollinating trees

Rainforest in
escarpment gullies

Rainforest patches
survive in unburnt areas,
including by rivers

River vegetation
fenced off from
cattle

On pastoral property,
country is managed
for sustainable beef
production

Fruit-eating birds
move between
rainforest patches
spreading seeds

Patchy fires prevent big
late Dry season fires,
protect rainforests and
create habitat mosaic
in savanna

Permanent
water hole

Seed-eating birds
rely on patchy fires to
find food all year

Waterbirds move between
wetlands seeking areas of right
depth for feeding and breeding

Healthy
mangroves

Deep
billabongs

Paperbark trees growing
around healthy wetlands

Natural flows maintain
productive fisheries

DEGRADED COUNTRY IN THE NORTH

Strip mine

Fruit-eating birds fail to reach patches of rainforest isolated by clearing, so seeds not spread

Big late Dry season fires damage country and affect many animals

Seasonal nectar supplies for wildlife reduced by clearing

Dam severely damages river health and reduces fish populations

Rainforest dries out as water table lowers

Water pumped from river, water table lowers and rainforest dries out

Cleared pastures

Irrigated paddocks

River vegetation and banks damaged by cattle

Water supplied from bore

Loss of wetlands reduces breeding opportunities for water birds

Overgrazing on fragile soils leads to erosion, causing sand to fill in waterholes

Pigs and buffalo damage wetlands and rainforests

Billabong gets flooded less

Weeds like Mimosa and Rubbervine choke wetlands and bushland

Sea level rises from global warming – and salt water kills coastal paperbark swamp

Changed fire practices reduce food for seed-eating birds

Reduced river flows lead to lower fish populations

APPENDIX 1 METHODS OF ANALYSES

CALCULATION OF GLOBAL LANDSCAPE INTEGRITY INDEX

by Brendan Mackey and Luciana Porfirio

Introduction

Our approach to calculating a global landscape integrity index for Tropical Savanna Woodland builds upon the method applied by Miles *et al.* (2006). All spatial analyses were undertaken using ARCMap Version 9.0 GIS (ESRI 2004). The index developed here combined three global data sets:

- A land cover classification generated from satellite imagery by the International Geosphere-Biosphere Programme (IGBP). This layer was used to identify landscapes containing savanna vegetation, and savanna that has been cleared for cropping;
- Estimates of livestock densities for cattle, sheep and goats produced by the Food and Agricultural Organisation of the United Nations (FAO); and
- Human population numbers produced by the Center for International Earth Science Information Network (CIESIN).

Analyses were restricted to data falling within the tropical zone, defined by the Tropic of Cancer and the Tropic of Capricorn. The three indices values for each grid cell were summed to give a relative index of Landscape Integrity. Details of the data sets and analytical techniques follow.

Global Land Cover data

Documentation about the IGBP Global Land Cover classification can be found at this URL: http://edcsns17.cr.usgs.gov/glcc/globdoc2_0.html. Our analyses used the Version 2.0 release of the global land cover characteristics database. This land cover classification was developed by the US Geological Survey's (USGS) National Center for Earth Resources Observation and Science (EROS), the University of Nebraska–Lincoln (UNL), and the Joint Research Centre of the European Commission (Loveland *et al.* 2000). The data set is derived from 1 km Advanced Very High Resolution Radiometer (AVHRR) data spanning a 12-month period (April 1992 – March 1993). The base data used are the International Geosphere Biosphere Programme (IGBP) 1 km AVHRR 10-day composites for April 1992 through March 1993 (Eidenshink and Faundeen 1994). Multitemporal AVHRR

NDVI data are used to divide the landscape into land cover regions, based on seasonality. A data quality evaluation was conducted and is reported by Zhu and Yang (1996).

For these analyses we followed the definition of Savanna as proposed by Scholes and Hall (1996) who argued that the importance of the tropical tree-grass systems has been obscured in the past by the vagueness of vegetation type definitions. They noted that the woodier savanna variations are often grouped with tropical forests, while the grassier forms are lumped with temperate grasslands. In reality, related vegetation types form a seamless continuum which can only be divided into distinct structural types by applying arbitrary limits. They recommended the following definitions, which we drew upon in defining savanna for the purposes or this study:

- *Forests* have complete tree canopy cover and three or more overlapping vegetation strata;
- *Woodlands* have 50–100% tree canopy cover by trees, and a sometimes sparse, but always significant gramineous layer;
- *Savannas* have 10–50% cover by woody plants, and in the unexploited state, a well-developed grass layer; and
- *Grasslands* have less than 10% tree cover.

Based on the Scholes and Hall definition, we mapped the global distribution of Savanna using the two relevant classes of the IGBP Land Cover classification data set, namely, class (8) Woody Savannas and class (9) Savannas. In addition, class (12) was used to map Savanna that had been converted to cropland, and class (14) to map land that is a mosaic of cropland/natural vegetation.

Global Livestock Data

The global livestock data were produced by FAO (2005). The data value units are densities (animals per square kilometre). They are corrected for the year 2000, and comprise a combination of observed and predicted estimates. Density data were available for eight species (cattle, buffalo, sheep, goats, pigs, chickens, bovines, small ruminants). We selected three indicative species, namely, cattle, sheep and goats. Note that the data did not include density estimates of feral livestock species.

Global Human Population Data

The gridded data for the world's population were in units of *persons per km^2* and were georeferenced at a 2.5 arc-minutes resolution.

These data were produced by the Center for International Earth Science Information Network (CIESIN), Columbia University; United Nations Food and Agriculture Programme (FAO); and Centro Internacional de Agricultura Tropical (CIAT). The data set were published in 2005 as 'Gridded Population of the World: Future Estimates (GPWFE). Palisades, NY and the Socioeconomic Data and Applications Center (SEDAC), Columbia University. The data are available at http://sedac.ciesin.columbia.edu/gpw.

Spatial Data Analysis

In preparation for analysis, the three data sets were converted to the same geographic projection and spatial resolution. The projection used to calculate the index was World Geodetic System spheroid of 1984 (WGS84), and the resolution was 1 km. Analyses were restricted to data falling within the tropical zone, defined by the Tropic of Cancer and the Tropic of Capricorn which lie at 23.5 degrees latitude north and south of the Equator. This tropical region was extended to 25 degrees north and south to provide an appropriate geographic buffer for the spatial analyses.

The gridded values of the Global Livestock Densities and Global Human Population Data were first normalised by their data range, resulting in relative indices scaled 0.0–1.0:
$$Xn = (X - Xmin) / (Xmax - Xmin)$$
where X is the data value in each grid cell for the GIS layer, $Xmin$ is the minimum cell value in the GIS layer, and $Xmax$ is the maximum cell value in the GIS layer.

This step produced a relative Livestock Index and a Human Population Index, where '1' corresponds with high levels and '0' low levels of livestock and humans.

The Savanna data layer produced from the IGBP Land Cover Classification was assigned relative index values as follows: 0.0 = Savannas; 0.5 = Cropland/Natural Vegetation Mosaic; 1.0 = Cropland

The three relative indices values for each grid cell (i.e., Livestock Index, Human Population Index, Savanna Index) were then summed to generate a new GIS layer with values ranging from 0.0–3.0. A final Landscape Integrity Index (*LI*) was then calculated:
$$LI = 1 - ((X - Xmin) / (Xmax - Xmin))$$
where X is the summed index value (Livestock,

Human Population, Savanna) for each grid cell in the GIS layer, *Xmin* is the minimum summed index value in the GIS layer, and *Xmax* is the maximum summed index value. Landscape Integrity Index values of 1.0 indicated locations that have not been exposed to high levels of the three threatening processes. The natural log was then taken of the Landscape Integrity Index layer as the distribution of the values was highly skewed with orders of magnitude difference in the value range.

References

Eidenshink, JC & Faundeen, JL 1994, 'The 1 km AVHRR Global Land Data Set: First Stages in Implementation', *International Journal of Remote Sensing* 15, pp. 17:3, 443–3,462.

ESRI 2004, ESRI® ArcMap™ 9.0, ESRI Inc.

FAO 2005, *Global Livestock Distributions*, data archive produced by Environmental Research Group Oxford Ltd for the Food and Agriculture Organisation of the United Nations, Rome, http://ergodd.zoo. ox.ac.uk and http://ergoonline.co.uk. (FAO were the funding organisation.)

Loveland, TR, Reed, BC, Brown, JF, Ohlen, DO, Zhu, J, Yang, L & Merchant, JW 2000, 'Development of a Global Land Cover Characteristics Database and IGBP DISCover from 1km AVHRR Data', *International Journal of Remote Sensing*, vol. 21, no. 6/7, pp. 1:303–1,330.

Miles, L, Newton, AC, Defries, RS, Ravilious, C, May, I, Blyth, S, Kapos, V & Gordon, JE 2006, 'A global overview of the conservation status of tropical dry forests', *Journal of Biogeography*, pp. 33:491–505.

Scholes, RJ & Hall, DO 1996, 'The carbon budget of tropical savannas, woodlands and grasslands', *Global Change: effects on coniferous forests and grasslands*, eds. A Breymeyer *et al.*, SCOPE 56, John Wiley, Chichester, pp. 66–100.

Zhu, L & Yang, L 1996, Characteristics of the 1 km AVHRR data set for North America. *International Journal of Remote Sensing*, www. informaworld.com/smpp/title~content= t713722504~db=all~tab=issueslist~bran ches=17#v17, 17 (10), pp. 1915–1924.

DATA AND METHODS FOR CONTINENTAL COMPARISON OF AUSTRALIAN WOODLANDS

by Brendan Mackey and Luciana Porfirio

Data

There were three geographic data sets used in the continental comparison of the condition of Australia's woodland vegetation: (1) Carnahan's map of pre-1788 Australian vegetation (Commonwealth of Australia 2003), recently modified with supplementary data (Nicholas Gellie, *pers. comm.*); hereafter the Carnahan map; (2) the Integrated Vegetation Cover V1 of the Bureau of Rural Sciences (BRS 2003); hereafter IVC1; and (3) An unpublished continental assessment of vegetation condition that used the VAST approach (Thackway and Lesslie 2006); hereafter, the VAST data.

The pre-1788 distribution of Australian Low and Medium Woodland was obtained from the Carnahan map at a map scale of 1:5 million scale map. The map broadly shows the probable state of Australia's vegetation around 1788 when European settlement began, depicting areas over 30,000 hectares as well as small areas of significant vegetation such as rainforest. Attribute information includes: growth form of tallest and lower stratum; foliage cover of tallest stratum; and dominant floristic types. The mapped data are available as a GIS layer, geo-coded with geographical coordinates (latitude and longitude) in decimal degrees using the Australian Geodetic Datum 1966 (AGD 1966).

The pre-1788 distribution of Australian Woodland vegetation was mapped by selecting the following formations from the Carnahan map:
1. Eucalypt Low Open Woodland
2. Eucalypt Low Woodland
3. Eucalypt Melaleuca Low Woodland
4. Eucalypt Open Woodland
5. Eucalypt Woodlands
6. Low Open Woodland
7. Low Woodland
8. Melaleuca Low Open Woodland
9. Melaleuca Low Woodland
10. Melaleuca Woodland
11. Terminalia/Lysiphyllum Low Open Woodland

Based on the these formation classes, we were able to distinguish between Low and Medium Woodlands where Low Woodland has a tree canopy <10 metres, and Medium Woodland has a tree canopy 10–30 metres.

Geographic data about land cleared for cropping and other forms of intensive land use activity were obtained from the IVC1 which was compiled from a number of recent vegetation-related datasets. Data are stored as a raster of 100 m resolution and are projected in Albers conic equal-area coordinates. Version 1 incorporates a selection of the latest available vegetation data as at July 2003. Vegetation cover in the IVC1 dataset is described using a 12-class attribute schema that was developed to meet vegetation-related information needs of the Commonwealth Government natural resource management arena.

The VAST data were generated following the approach documented by Thackway and Lesslie (2006) using available continental scaled data. This new continental VAST analysis was made available curtsey of the authors. VAST is an approach to classifying vegetation condition along a set of ranked (ordinal) classes as detailed in Thackway and Lesslie (2006).

Calculations

The spatial data were stored in the D_WGS_1984 (Degrees) map projection. However, for the GIS calculation, the data were re-projected to Lambert Conformal Conic (Meters). The cell resolution used for the analysis was 0.01°, which is the resolution of the VAST data grid. ArcMap was used to generate the maps and calculate the geographic statistics. The pre-1788 Woodland coverage was the overlaid with (1) land cleared for intensive land use (including cropping) from IVC1 and (2) the continental VAST data. The percentage of each VAST class (and land cleared for cropping) was calculated for the pre-1788 distribution of Low and Medium Woodland. These percentages were then calculated for Northern Australia.

References

Australian Survey and Land Information Group (AUSLIG) 1990, *Atlas of Australian Resources, Vegetation,* third series, vol. 6, Canberra, viewed 20 July 2006, Australian Government Geoscience Australia, 2004, http://www. ga.gov.au/nmd/products/thematic/veg.htm.

BRS, The Bureau of Rural Sciences 2003, *Integrated Vegetation Cover V1,* Australian Government Department of Agriculture, Fisheries and Forestry, viewed 20 July 2006, ivc03_md.doc (D:\Data\wc_Data\ vegetation\BRS\Integrated_veg\vcov_exp).

Thackway, R & Lesslie, R 2006, 'Reporting vegetation condition using the Vegetation Assets, States and Transitions (VAST) framework', *Ecological Management & Restoration* 7, pp. 1: 53–62.

AUSTRALIAN RIVER DISTURBANCE INDEX

The River Disturbance Index (RDI) shown in Figure 4.6 is an update of the continental Wild Rivers analysis undertaken by Janet Stein and colleagues and reported in Stein *et al.* (2002). The analysis shown in Map 1 was kindly provided for this report by Janet Stein. The RDI is designed to assess anthropogenic river disturbance. The GIS-based method calculates indices of disturbance for individual stream segments. The method considers both point and diffuse impacts on water quality, and includes both catchment and in-stream factors that alter flow regimes. Specifically, the method used data about the intensity and extent of human activities within the catchment, along with in-stream structures that alter the flow regime.

References

Stein, JL, Stein, JA & Nix, HA 2002, 'Spatial analysis of anthropogenic river disturbance at regional and continental scales: identifying the wild rivers of Australia', *Landscape and Urban Planning,* pp. 60:1–25.

REFERENCES

Abbott, I & Burbidge, AA 1995, 'The occurrence of mammal species on the islands of Australia: a summary of existing knowledge', *CALMScience* 1, pp. 259–324.

Abrahams, H, Mulvaney, M, Glasco, D & Bugg, A 1995, *Areas of conservation significance on Cape York Peninsula*, Australian Heritage Commission, Canberra.

Andersen, AN 1991, 'Responses of ground-foraging ant communities to three experimental fire regimes in a savanna forest of tropical Australia', *Biotropica* 23, pp. 575–585.

Andersen, AN 1992, 'Regulation of "momentary" diversity by dominant species in exceptionally rich ant communities of the Australian seasonal tropics', *American Naturalist* 140, pp. 401–420.

Andersen, AN 1993a 'Ant communities in the Gulf region of Australia's semi-arid tropics: species composition, patterns of organization and biogeography', *Australian Journal of Zoology* 41, pp. 399–414.

Andersen, AN 1993b 'Ants as indicators of restoration success at a uranium mine in tropical Australia', *Restoration Ecology* 1, pp. 156–167.

Andersen, AN & Lonsdale, WM 1990, 'Herbivory by Insects in Australian Tropical Savannas: A Review', *Journal of Biogeography*, vol. 17, No. 4/5, *Savanna Ecology and Management: Australian Perspectives and Intercontinental Comparisons July–September*, pp. 433–444.

Andersen, AN & Patel, AD 1994, 'Meat ants as dominant members of Australian ant communities: an experimental test of their influence on the foraging success and forager abundance of other species', *Oecologia* 98, pp. 15–24.

Andersen, AN 2000, *The Ants of Northern Australia: A Guide to the Monsoonal Fauna*, CSIRO Publishing, Collingwood.

Andersen, AN, Azcarate, FM & Cowie, ID 2000a), 'Seed selection by an exceptionally rich community of harvester ants in the Australian seasonal tropics', *Journal of Animal Ecology* 69, pp. 975–984.

Andersen, AN, Lowe, LM & Rentz, DCF 2000b), 'The grasshopper (Orthoptera: Acridoidea, Eumastacoidea and Tettigonioidea) fauna of Kakadu National Park in the Australian seasonal tropics: biogeography, habitat associations and functional groups', *Australian Journal of Zoology* 48, pp. 431–442.

Andersen, AN, Cook, GD & Williams, RJ (eds) 2003, *Fire in tropical savannas: the Kapalga experiment*, Springer, New York.

Andersen, AN & Majer, JD 2004, 'Using invertebrates as bioindicators in land management: ants show the way Down-Under', *Frontiers in Ecology and the Environment* 2, pp. 291–298.

Andersen, AN, Woinarski, JCZ & Hoffmann, BD 2004, 'Biogeography of the ant fauna of the Tiwi islands, in northern Australia's monsoonal tropics', *Australian Journal of Zoology* 52, pp. 97–110.

Andersen, AN, Hertog, T & Woinarski, JCZ 2006, 'Long-term fire exclusion and ant community structure in an Australian tropical savanna: congruence with vegetation succession', *Journal of Biogeography* 33, pp. 823–832.

Andrew, MH & JJ Mott, 'Annuals with transient seed banks: the population biology of indigenous *Sorghum* species of tropical north-west Australia', *Australian Journal of Ecology*, 1983, pp. 8:265–276.

Anon. 2006, 'Baby boabs: the exciting new taste sensation from the Kimberly in Western Australia', *Agriculture WA Bulletin* 4672, ISSN 1448-0352, pp. 1–13.

Archer, M, Hand, SJ & Godthelp, H 1991, *Riversleigh: the story of animals in ancient rainforests of inland Australia*', Reed, Sydney.

Australian Government, *State of the Environment Report 2001*.

Australian Government, *State of the Environment Report 2006*.

Australian Survey and Land Information Group (AUSLIG) 1990, *Atlas of Australian Resources*, vol. 6, *Vegetation*, third series, Canberra, viewed 20 July 2006, Australian Government Geoscience Australia, 2004, http://www.ga.gov.au/nmd/products/thematic/veg.htm.

Banfai, DS & Bowman, DMJS 2005, 'Dynamics of a savanna-forest mosaic in the Australian monsoon tropics inferred from stand structures and historical aerial photography', *Australian Journal of Botany* 53, pp. 185–194.

Banfai, DS & Bowman, DMJS 2006, 'Forty years of lowland monsoon rainforest expansion in Kakadu National park, Northern Australia', *Biological Conservation* 131, pp. 553–565.

Bauer, FH (ed.) 1977, *Cropping in North Australia: Anatomy of Success and Failure*, Australian National University, Canberra.

Baum, DA, Small, RL & Wendel, JF 1998, 'Biogeography and floral evolution of baobabs (*Adansonia, Bambaceae*) as inferred from multiple data sets', *Systematic Biology* 47, pp. 2:181–207.

Bayliss, BL, Brennan, KG, Eliot, I Finlayson, CM, Hall, RN, House, T, Pidgeon, RWJ, Walden, D & Waterman, P 1998, 'Vulnerability assessment of predicted climate change and sea level rise in the Alligator Rivers region, Northern Territory, Australia', *Supervising Scientist Report* 123, Supervising Scientist, Canberra.

Bayon, R, Hawn, A & Hamilton, K (eds) 2004, *Voluntary carbon markets: an international business guide to what they are and how they work*, Sterling VA: Earthscan, London.

Beard JS, Chapman AR, Gioia P 2000, 'Species richness and endemism in the Western Australian flora', *Journal of Biogeography* 27, pp. 1257–1268.

Begg, GW, van Dam, RA, Lowry, JB, Finlayson, CM & Walden, DJ 2001, 'Inventory and risk assessment of water dependent ecosystems in the Daly Basin, Northern Territory, Australia', *Supervising Scientist Report* 162, Supervising Scientist, Canberra.

Blandford, DC 1979, *Ord River catchment (NT) regeneration and research*, Territory Parks and Wildlife Commission, Darwin.

Boekel, C 1979, 'Notes on the status and behaviour of the purple-crowned fairy-wren *Malurus coronatus* in the Victoria River Downs area, Northern Territory', *Australian Bird Watcher* 8, pp. 91–97.

Bowman DMJS, Wilson BA, Dunlop CR 1988, 'Preliminary biogeographic analysis of the Northern Territory vascular flora', *Australian Journal of Botany* 36, pp. 503–517.

Bowman DMJS, Panton WJ 1993, 'Decline of *Callitris intratropica* RT Baker & HG Smith in the Northern Territory: implications for pre- and post-European colonization fire regimes', *Journal of Biogeography* 20, pp. 373–381.

Bowman, DMJS 1997, 'Observations on the demography of the Australian Boab (*Adansonia gibbosa*) in the North-west of the Northern Territory, Australia', *Australian Journal of Botany* 45, pp. 893–904.

Bowman, DMJS, Walsh, A & Milne, DJ 2001a, Forest expansion and grassland contraction within a Eucalyptus savanna matrix between 1941 and 1994 at Litchfield National Park in the Australian monsoon tropics', *Global Ecology and Biogeography* 10, pp. 535–548.

Bowman DMJS, Price O, Whitehead PJ, Walsh, A 2001b, 'The "wilderness effect" and the decline of *Callitris intratropica* on the Arnhem Land Plateau, Northern Australia', *Australian Journal of Botany* 49, pp. 665–672.

Brennan, K 1996, 'An annotated checklist of the vascular plants of the Alligator Rivers Region, Northern Territory, Australia', *Supervising Scientist Report* 109, Supervising Scientist, Canberra.

BRS, The Bureau of Rural Sciences 2003, *Integrated Vegetation Cover V1*, Australian Government Department of Agriculture, Fisheries and Forestry, viewed 20 July 2006, ivc03_md.doc (D:\Data\wc_Data\vegetation\BRS\Integrated_veg\vcov_exp).

Burbidge, AA, McKenzie, NL & Kenneally, KF 1991, *Nature Conservation Reserves in the Kimberley Western Australia*, Department of Conservation and Land Management.

Burnett, S 1997, 'Colonising cane toads cause population declines in native predators: reliable anecdotal information and management implications', *Pacific Conservation Biology* 3, pp. 65–72.

Chapman, S 2000, 'Occurrence and eradication of a small population of the Eurasian Tree Sparrow *Passer montanus* in Darwin', *Northern Territory Naturalist* 16, pp. 32–34.

Chappell, J & Thom, BG 1986, 'Coastal morphodynamics in North Australia: review and prospect', *Australian Geographical Studies* 24, pp. 110–127.

Chen, X, Hutley, LB & Eamus, D 2003, 'Carbon balance of a tropical savanna of northern Australia', *Oecologia* 137, pp. 405–416.

Clarkson JR, Kenneally KF 1988, 'The floras of Cape York and the Kimberley: a preliminary comparative analysis', *Proceedings of the Ecological Society of Australia* 15, pp. 259–266.

Cook, GD, Setterfield, SA & Maddison, J 1996, 'Scrub invasion of a tropical wetland: implications for weed management', *Ecological Applications* 6, pp. 531–537.

Cook, GD, Liedloff, AC, Eager, RW, Chen, X, Williams, RJ, O'Grady, AP & Hutley, LB 2005, 'The estimation of carbon budgets of frequently burnt tree stands in savannas of northern Australia, using allometric analysis and isotopic discrimination', *Australian Journal of Botany* 53, pp. 621–630.

Cook, GD, Taylor, RJ, Williams, RJ & Banks, JCG 2005, 'Sustainable harvesting rates of ironwood, *Erythrophleum chlorostachys*, in the Northern Territory, Australia', *Australian Journal of Botany* 53, pp. 821–826.

Cook, GD & Dias, L 2006, 'It was no accident: deliberate plant introductions by Australian government agencies during the 20th century', *Australian Journal of Botany* 54, pp. 601–625.

Coombs, HC 1977, 'Opening remarks, in F H Bauer (ed), *Cropping in North Australia: Anatomy of Success and Failure*', Australian National University, Canberra, pp. 8–9.

Cork, S, Sattler, P & Alexandra, J 2006, 'Biodiversity' theme commentary prepared for the 2006 Australian State of the Environment Committee, Department of Environment and Heritage, Canberra.

Cowie, ID & Werner, PA 1993, 'Alien plant species invasive in Kakadu National park, tropical northern Australia', *Biological Conservation* 63, pp. 127–135.

Crowley, GM & Garnett, ST 1998, 'Vegetation change in the grasslands and grassy woodlands of east-central Cape York Peninsula, Australia', *Pacific Conservation Biology* 4, pp. 132–48.

Crowley, GM & Garnett, ST 1999, 'Seed of the annual grasses *Schizachyrium* spp. as a food for tropical granivorous birds', *Australian Journal of Ecology*, pp. 24:208–220.

Crowley, GM & Garnett, ST 2000, 'Changing fire management in the pastoral lands of Cape York Peninsula of northeast Australia, 1623 to 1996', *Australian Geographic Studies* 38, pp. 10–26.

Crowley, GM 2001, 'Grasslands of Cape York Peninsula – a fire-dependent habitat', *Savanna Burning – Understanding and Using Fire in Northern Australia*, eds Dyer, R, Jacklyn, P, Partridge, I, Russell-Smith, J & Williams, R, Tropical Savannas CRC, Darwin, pp. 34.

Crowley, GM & Garnett, ST 2001, 'Growth, seed production and effect of defoliation in an early flowering perennial grass, *Alloteropsis semialata* (Poaceae), on Cape York Peninsula, Australia', *Australian Journal of Botany*, pp. 49:735–43.

Crowley, GM, Garnett, ST & Shephard, S 2004, *Management guidelines for golden-shouldered parrot conservation*, Queensland Parks and Wildlife Service, Brisbane.

CSIRO 1953, 'Survey of Katherine-Darwin Region, 1946', *Land Research series* No. 1, CSIRO, Melbourne.

Dahl, K 1897, 'Biological notes on north-Australian mammalia', *Zoologist*, Series 4, 1, pp. 189–216.

Dahl, K 1926, *In savage Australia: an account of a hunting and collecting expedition to Arnhem Land and Dampier Land*, Philip Allan & Co., London.

Doody, JS, Green, B, Sims, R, Rhind, D, West, P & Steer, D 2006, 'Indirect impacts of invasive cane toads *Bufo marinus* on nest predation in pig-nosed turtles *Carettochelys insculpta*', *Wildlife Research* 33, pp. 349–354.

Dostine, PL, Johnson, GC, Franklin, DC & Hempel, C 2001, 'Seasonal use of savanna landscapes by the Gouldian finch, *Erythrura gouldiae* and co-existing finches in the Yinberrie Hills area, Northern Territory', *Wildlife Research* 28, pp. 445–458.

Dostine, PL 2002, *Australia-wide assessment of river health: Northern Territory program*, NT Department of Infrastructure Planning and Environment, Darwin.

Doupe RG & Pettit NE 2002, 'Ecological perspectives on regulation and water allocation for the Ord River, Western Australia', *River Research and Applications* 18, pp. 307–320.

Duretto MF 1997, 'Taxonomic notes on *Boronia* species of north-western Australia, including a revision of the *Boronia lanuginosa* group (*Boronia* section *Valvatae*: Rutaceae)', *Nuytsia* 11, pp. 301–346.

Duretto MF & Ladiges PY 1997, 'Morphological variation within the *Boronia grandisepala* group (Rutaceae) and the description of nine taxa endemic to the Northern Territory, Australia', *Australian Systematic Botany* 10, pp. 249–302.

du Toit, JT, Walker, BH & Campbell, BM 2003, 'Conserving tropical nature: current challenges for ecologists', *Trends in Ecology and Evolution*.

Edwards, A, Kennett, R, Price, O, Russell-Smith, J, Spiers, G & Woinarski, J 2003, 'Monitoring the impacts of fire regimes on biodiversity in northern Australia: an example from Kakadu National Park', *International Journal of Wildland Fire* 12, pp. 427–440.

Eliot, I, Finlayson, CM & Waterman, P 1999, 'Predicted climate change, sea level rise and wetland management in the Australian wet-dry tropics', *Wetlands Ecology and Management* 7, pp. 63–81.

Fairfax, RJ & Fensham, RJ 2000, 'The effect of exotic pasture development on floristic diversity in central Queensland, Australia', *Biological Conservation* 94, pp. 11–21.

Fensham, RJ & Fairfax, RJ 2003, 'Assessing woody vegetation cover change in north-west Australian savanna using aerial photography', *International Journal of Wildland Fire* 12, pp. 359–367.

Fensham, RJ & Holman, JE 1999, 'Temporal and spatial patterns in drought-related tree dieback in Australian savanna', *Journal of Applied Ecology* 36, pp. 1035–1050.

Finlayson, CM, Lowry, J, Bellio, MG, Nou, S, Pidgeon, R, Walden, D, Humphrey, C & Fox, G 2006, 'Biodiversity of the wetlands of the Kakadu Region, northern Australia', *Aquatic Science* 68, pp. 374–399.

Firth, RSC, Woinarski, JCZ & Noske, RA 2006a, 'Home range and den characteristics of the brush-tailed rabbit-rat *Conilurus penicillatus* in the monsoonal tropics of the Northern Territory, Australia', *Wildlife Research* 33, pp. 397–408.

Firth, RSC, Woinarski, JCZ, Brennan, KG & Hempel, C 2006b, 'Environmental relationships of the brush-tailed rabbit-rat *Conilurus penicillatus* and other small mammals on the Tiwi Islands, Northern Australia', *Journal of Biogeography* 33, pp. 1820–1837.

Fisher, A 2001, *Biogeography and conservation of Mitchell grasslands in northern Australia*, PhD thesis, Northern Territory University, Darwin.

Ford, HA, Barrett, GW, Saunders, DA & Recher HF 2001, 'Why have birds in the woodlands of Southern Australia declined?', *Biological Conservation* 97 (1), pp. 71– 88.

Franklin, DC 1999, 'Opportunistic nectarivory: an annual dry season phenomenon among birds in monsoonal northern Australia', *Emu* 99, pp. 135–141.

Franklin, DC, Whitehead, PJ, Pardon, G, Matthews, J, McMahon, P & McIntyre, D 2005, 'Geographic patterns and correlates of the decline of granivorous birds in northern Australia', *Wildlife Research* 32, pp. 399–408.

Franklin, DC, Petty, AM, Williamson, GJ, Brook, BW & Bowman, DMJS, 2007 (in press), 'Monitoring contrasting land management in the savanna landscapes of Northern Australia', *Environmental Management*.

Fraser FJ 2001, *The impacts of fire and grazing on the Partridge Pigeon: the ecological requirements of a declining tropical granivore*, PhD thesis, Australian National University, Canberra.

Freeland, WJ, Winter, JW & Raskin, S 1988, 'Australian rock-mammals: a phenomenon of the seasonally dry tropics', *Biotropica* 20, pp. 70–79.

Garnett, ST & Crowley, GM 1994, 'Wet season feeding by four species of granivorous birds in the Northern Territory', *Australian Bird Watcher*, pp. 5:306–9.

Garnett, ST & Crowley, GM 1995, 'Feeding ecology of hooded parrots *Psephotus dissimilis* during the early wet season', *Emu*, pp. 95:54–61.

Garnett, ST & Crowley, GM 2003, 'National recovery plan for the golden-shouldered parrot *Psephotus chrysopterygius* 2003–2007', Queensland Environmental Protection Agency, Brisbane.

Garnett, ST & Crowley, GM 2003, 'Cape York Peninsula', *Wilderness: Earth's Last Wild Places*, eds RA Mittermeier, CG Mittermeier, PR Gil, J Pilgrim, G Fonesca, T Brooks & WR Konstant, Cemex, Mexico City, pp. 220–229.

Garnett, ST & Sithole, B 2007, *Sustainable Northern Landscapes and the Nexus with Indigenous Health: Healthy Country Healthy People: Final Draft*, Land and Water Australia, Canberra.

Gilligan, B 2006, *The Indigenous Protected Areas Programme*, 2006 evaluation, Department of Environment and Heritage, Canberra.

Graham, G & McKenzie, NL 2004, *A Conservation Case Study of Western Australia's Mitchell Subregion (North Kimberley 1) in 2003*, Department of Conservation and Land Management, Perth.

Grice, AC 2000, 'Weed management in Australian rangelands', *Australian weed management systems*, ed. BM Sindel, pp. 429–458. (RG & FJ Richardson, Frankston.)

Grice, AC 2006, 'The impacts of invasive plant species on the biodiversity of Australian rangelands', *Rangeland Journal* 28, pp. 27–36.

Grigg, GC, Johansen, K, Harlow, P & Taplin, LE 1986, 'Facultative aestivation in a tropical freshwater turtle *Chelodina rugosa*', *Comparative Biochemistry and Physiology* 83A, pp. 321–323.

Groves, RH, Hosking, JR, Batianoff, GN, Cooke, DA, Cowie, ID, Johnson, RW, Keighery, GJ, Lepschi, BJ, Mitchell, AA, Moerkerk, M, Randall, RP, Rozefields, AC, Walsh, NG & Waterhouse, BM 2003, *Weed categories for natural and agricultural ecosystem management*, Bureau of Rural Sciences, Canberra.

Hoffmann, BD, Andersen, AN & Hill, GJE 1999, 'Impact of an introduced ant on native rain forest invertebrates *Pheidole megacephala* in monsoonal Australia', *Oecologia* 120, pp. 595–604.

Hoffmann, BD 2000, 'Changes in ant species composition and community organisation along grazing gradients in semi-arid rangelands of the Northern Territory', *The Rangeland Journal* 22, pp. 171–189.

Hoffmann, BD, Griffiths, AD & Andersen, AN 2000, 'Response of ant communities to dry sulphur deposition from mining emissions in semi-arid northern Australia, with implications for the use of functional groups', *Austral Ecology* 25, pp. 653–663.

Hoffmann, BD & O'Connor, S 2004, 'Eradication of two exotic ants from Kakadu National Park', *Ecological management & Restoration* 5, pp. 98–105.

Jacklyn, PM 1992, 'Magnetic termite mound surfaces are oriented to suit wind and shade conditions', *Oecologia* 91, pp. 385–395.

Johnson, C 2006, *Australia's mammal extinctions: a 50,000 year history*, Cambridge University Press, Cambridge.

Johnson, C, Isaacs, JL & Fisher, DO 2006, 'Rarity of a top predator triggers continent-wide collapse of mammal prey: dingoes and marsupials in Australia', *Proceedings of the Royal Society B* xxx.

Johnston, FH, Kavanagh, AM, Bowman, DMJS & Scott, RK 2002, 'Exposure to bushfire smoke and asthma: an ecological study', *Medical Journal of Australia* 176, pp. 535–538.

Johnston, F, Jacklyn, S, Vickery, A & Bowman, DMJS 2007, 'Ecohealth and the Aboriginal testimony of the nexus between human health and place', *Ecohealth*, accepted June 2007

Johnson, KA & Kerle, JA 1991, *Flora and vertebrate fauna of the Sir Edward Pellew group of islands, Northern Territory*, report to the Australian Heritage Commission, Conservation Commission of the Northern Territory, Alice Springs.

Kennett, R & Christian, K 1993, 'Aestivation by freshwater crocodiles *Crocodylus johnstoni* occupying a seasonally ephemeral creek in tropical Australia', *Herpetology in Australia: a diverse discipline*, eds D Lunney and D Ayers, Surrey Beatty & Sons: Chipping Norton, pp. 315–319.

Kingsford, RT 2000, 'Review: Ecological impacts of dams, water diversions and river management on floodplain wetlands in Australia', *Austral Ecology* (25), pp. 109–127.

Land and Water Resources Australia, Land and Water Resources Audit, *Assessment of Terrestrial Biodiversity 2002*.

Lees, BG 1992, 'Geomorphological evidence for late Holocene climatic changes in northern Australia', *Australian Geographer*, pp. 23, 1–10.

Letts, GA, Bassingthwaighte, A & de Vos, WEL 1979, *Feral animals in the Northern Territory*, Report of Board of Inquiry, Department of Primary Production, Darwin.

Lewis, D 2002, *Slower than the eye can see. Environmental change in northern Australia's cattle lands. A case study from the Victoria River District, Northern Territory*, Tropical Savannas CRC, Darwin.

Mackey, BG, Nix, H & Hitchcock, P 2001, *The National Heritage significance of Cape York Peninsula*, ANUTECH, Canberra.

Mackey, BG, Soulé, ME, Nix, HA, Recher, HF, Lesslie, RG, Williams, JE, Woinarski, JCZ, Hobbs, J & Possingham, HP 2007, 'Towards a scientific framework for the WildCountry project', *Key Topics and Perspectives in Landscape Ecology*, Chapter 11, eds J Wu and RJ Hobbs, Cambridge University Press, pp. 92–208.

Macknight, CC 1976, *The voyage to Marege*, Melbourne University Press, Melbourne.

Madsen, T & Shine, R 1996, 'Seasonal migration of predators and prey – a study of pythons and rats in tropical Australia', *Ecology* 77, pp. 149–156.

Madsen, T, Ujvari, B, Shine, R, Buttemer, W & Olsson, M 2006, 'Size matters: extraordinary rodent abundance on an Australian tropical flood plain', *Austral Ecology* 31, pp. 361–365.

Majer, JD 1984, 'Recolonisation by ants in rehabilitated open-cut mines in northern Australia', *Reclamation and Revegetation Research* 2, pp. 279–298.

Majer, JD 1990, 'The abundance and diversity of arboreal ants in northern Australia', *Biotropica*, pp. 22:191–199.

Martin, TG, Campbell, SD & Grounds, S 2006, 'Weeds of Australian rangelands', *Rangeland Journal* 28, pp. 3–26.

McKenzie, NL, Burbidge, AA, Chapman, A & Youngson K 1978. 'Part III Mammals: The islands of the north-west Kimberley, Western Australia', *Wildlife Research Bulletin Western Australia* No 7, eds AA Burbidge & NL McKenzie, pp. 22–28.

McKenzie, NL 1981.'Mammals of the Phanerozoic south-west Kimberley, Western Australia: biogeography and recent changes', *Journal of Biogeography* 8, pp. 263–280.

McKenzie, NL, Johnston, RB & Kendrick, PG (eds) 1991, *Kimberley rainforests of Australia*, Surrey Beatty & Sons, Chipping Norton.

Mott, JJ 1978, 'Dormancy and germination in five native grass species from savannah woodland communities of the Northern Territory', *Australian Journal of Botany*, pp. 26:621–631.

Natural Resources Policies and Programs Committee Biodiversity Working Group 2005, *A national approach to biodiversity decline*, report to the Natural Resources Management Ministerial Council.

Neldner, VJ & Clarkson, JR 1995, *Vegetation survey and mapping of Cape York Peninsula*, CYPLUS, Brisbane.

Neldner, VJ, Fensham, RJ, Clarkson, JR & Stanton, JP 1997, 'The natural grasslands of Cape York Peninsula, Australia: Description, distribution and conservation status', *Biological Conservation* 81, pp. 121–136.

Newsome, AE, Catling, PC, Cooke, BD & Smyth, R 2001, 'Two ecological universes separated by the dingo barrier fence in semi-arid Australia: interactions between landscapes, herbivory and carnivory, with and without dingoes', *Rangeland Journal* 23, pp. 71–98.

NGIS Australia 2004, *Australia's Tropical Rivers – Data Audit*, Land & Water Australia, Canberra. Product Number PR040674.

Nix, HA & Kalma, JD 1972, 'Climate as a dominant control in the biogeography of northern Australia and New Guinea', *Bridge and Barrier: the natural and cultural history of Torres Strait*, ed. D Walker, ANU, Canberra.

Nix, HA 2004, 'Inverting the paradigm', *Pacific Conservation Biology* 10, pp. 76.

Noirot, CH 1970, 'The Nests of Termites', *Biology of Termites* vol. 2, eds K Krishna & FM Weesner, Academic Press: New York and London, pp. 73–125.

Ockwell, D & Lovett, JC 2005, 'Fire assisted pastoralism vs. sustainable forestry – the implications of missing markets for carbon in determining optimal land use in the wet-dry tropics of Australia', *Journal of Environmental Management* 75, pp. 1–9.

Palmer, C & Woinarski, JCZ 1999, 'Seasonal roosts and foraging movements of the Black Flying-fox *Pteropus alecto* in the Northern Territory: resource tracking in a landscape mosaic', *Wildlife Research* 26, pp. 823–838.

Pardon, LG, Brook, BW, Griffiths, AD & Braithwaite, RW 2003, 'Determinants of survival for the northern brown bandicoot under a landscape-scale fire experiment', *Journal of Animal Ecology* 72, pp. 106–115.

Pople, AR, Grigg, GC, Cairns, SC, Beard, LA & Alexander, P 2000, 'Trends in the numbers of red kangaroos and emus on either side of the South Australian dingo fence: evidence for predator regulation?', *Wildlife Research* 27, pp. 269–276.

Price, O & Bowman, DMJS 1994, 'Fire-stick forestry: a matrix model in support of skilful fire management of *Callitris intratropica* RT Baker by north Australian Aborigines', *Journal of Biogeography* 21, pp. 573–580.

Price, O, Milne, D, Connors, G, Harwood, B, Woinarski, JCZ & Butler, M 2000, *A conservation plan for the Daly Basin bioregion*, report to Natural Heritage Trust, Parks and Wildlife Commission of the Northern Territory: Darwin.

Pringle, H & Tinley, K 2003, 'Are we overlooking critical geomorphic determinants of landscape change in Australian rangelands?', *Ecological Management & Restoration*, vol. 4, issue 3, pp. 180.

Prior, LD, Bowman, DMJS & Brook, BW 2007, 'Growth and survival of two north Australian relictual tree species, *Allosyncarpia ternata* (Myrtaceae) and *Callitris intratropica* (Cupressaceae)', *Ecological Review* 22, pp. 228–236.

Reichel, H & Andersen, AN 1996, 'The rainforest ant fauna of Australia's Northern Territory', *Australian Journal of Zoology* 44, pp. 81–95.

Robinson, CJ, Smyth, D & Whitehead, PJ 2005, 'Bush tucker, bush pets and bush threats: cooperative management of feral animals in Australia's Kakadu National Park', *Conservation Biology* 19, pp. 1385–1391.

Rogers *et al.* 2003, *Life along land's edge – wildlife on the shores of Roebuck Bay, Broome*, CALM, Perth.

Rossiter, NA, Setterfield, SA, Douglas, MM & Hutley, LB 2003, 'Testing the grass-fire cycle: alien grass invasion in the tropical savannas of northern Australia', *Diversity and Distributions* 9, pp. 169–176.

Rowley, I 1993, 'The purple-crowned fairy-wren *Malurus coronatus*, History, distribution and present status', *Emu* 93, pp. 220–234.

Russell-Smith, J & Bowman, DMJS 1992, 'Conservation of monsoon rainforest isolates in the Northern Territory, Australia', *Biological Conservation* 59, pp. 51–63.

Russell-Smith, J, McKenzie, NL & Woinarski, JCZ 1992, 'Conserving vulnerable habitat in northern and northwestern Australia: the rainforest archipelago', *Conservation and development issues in northern Australia*, eds. I Moffatt & AWebb, North Australia Research Unit, Darwin, pp. 63–68.

Russell-Smith, J, Lucas, D, Gapindi, M, Gunbunuka, B, Kapirigi, N, Namingum, G, Lucas, K, Giuliani, P & Chaloupka, G 1997, 'Aboriginal resource utilization and fire management practice in western Arnhem Land, monsoonal northern Australia: notes for prehistory and lessons for the future', *Human Ecology* 25, pp. 159–195.

Russell-Smith, J, Ryan, PG, Klessa, D, Waight, G & Harwood, R 1998, 'Fire regimes, fire-sensitive vegetation and fire management of the sandstone Arnhem plateau, monsoonal northern Australia', *Journal of Applied Ecology* 35, pp. 829–846.

Russell-Smith, J, Ryan, PG & Cheal, D 2002, 'Fire regimes and the conservation of sandstone heath in monsoonal northern Australia: frequency, interval, patchiness', *Biological Conservation* 104, pp. 91–106.

Russell-Smith, J, Whitehead, PJ, Cook, GD & Hoare, JL 2003, 'Response of *Eucalyptus*-dominated savann to frequent fires: lessons from Munmarlary 1973–1996', *Ecological Monographs* 73, pp. 349–375.

Russell-Smith, J, Stanton, PJ, Whitehead, PJ & Edwards. A 2004a, 'Rain forest invasion of eucalypt-dominated woodland savanna, Iron Range, north-eastern Australia: I successional processes', *Journal of Biogeography* 31, pp. 1293–1303.

Russell-Smith, J, Stanton, PJ, Edwards, A & Whitehead, PJ 2004b, 'Rain forest invasion of eucalypt-dominated woodland savanna, Iron Range, north-eastern Australia: II rates of landscape change', *Journal of Biogeography* 31, pp. 1293–1303.

Russell-Smith, J 2006, 'Recruitment dynamics of the long-lived obligate seeders *Callitris intratropica* (Cupressaceae) and *Petraeomyrtus punicea* (Myrtaceae)', *Australian Journal of Botany* 54, pp. 479–485.

Sattler, P & Creighton, C 2002, *Australian Terrestrial Biodiversity Assessment 2002*, National Land & Water Resources Audit, Land and Water Australia, Commonwealth of Australia.

Sattler, P & Glanznig, A 2006, *Building nature's safety net: a review of Australia's terrestrial protected area system, 1991–2004*, WWF, Sydney.

Schulmeister, J 1992, 'A Holocence pellen record from lowland tropical Australia', *The Holocene* 2, pp. 107–116.

Sebastien, L, Cook, PG, O'Grady, A & Eamus, D 2005, 'Groundwater used by vegetation in a tropical savanna riparian zone (Daly River, Australia)', *Journal of Hydrology*, pp. 310:280–293.

Sinden, J, Jones, R, Hester, S, Odom, D, Kalisch, C, James, R & Cacho, O 2004, 'The economic impact of weeds in Australia', *CRC for Australian Weed Management Technical Series* 8, pp. 1–55.

Skeat, AJ, East, TJ & Corbett, LK 1996, 'Impact of feral water buffalo', *Landscape and vegetation ecology of the Kakadu region, Northern Australia*, eds CM Finlayson and I von Oertzen, Kluwer, Dordrecht, pp. 155–178

Smith, JG & Phillips, BL 2006, 'Toxic tucker: the potential impact of cane toads on Australian reptiles', *Pacific Conservation Biology* 12, pp. 40–49.

Smith, LA & Johnstone, RE 1977, 'Status of the purple-crowned wren *Malurus coronatus* and buff-sided robin *Poecilodryas superciliosa* in Western Australia', *Western Australian Naturalist* 13, pp. 185–188.

Soulé, ME & Terborgh, J (eds) 1999, *Continental conservation: scientific foundations of regional reserve networks*, Island Press, Washington DC.

Soulé, ME, Mackey, BG, Recher, HF, Williams, JE, Woinarski, JCZ, Driscoll, D, Dennison, WC & Jones, M 2004, 'Continental connectivity: its role in Australian conservation', *Pacific Conservation Biology* 10, pp. 266–279.

Southgate, R, Palmer, C, Adams, M, Masters, P, Triggs, B & Woinarski, J 1996, 'Population and habitat characteristics of the Golden Bandicoot *Isoodon auratus* on Marchinbar Island, Northern Territory', *Wildlife Research* 23, pp. 647–664.

Specht, RL 1981, 'Major vegetation formations in Australia', *Ecological Biogeography of Australia*, ed. A Keast, Junk, The Hague, pp.163–298.

Stanton, P & Morgan, MG 1976, *Project RAKES: A Rapid Appraisal of Key and Endangered Sites – Report No. 1, the Queensland Case Study*, Department of Environment, Housing and Community Development, Canberra.

Start, AN, Burbidge, AA, McKenzie, NL & Palmer, C (in press) 'The status of mammals in the North Kimberley, Western Australia', *Australian Mammalogy*.

Stein, JL, Stein, JA & Nix, HA 2002, 'Spatial analysis of anthropogenic river disturbance at regional and continental scales: identifying the wild rivers of Australia', *Landscape and Urban Planning*, pp. 60:1–25.

Tassicker, AL, Kutt, AS, Vanderduys, E & Mangru, S 2006, 'The effects of vegetation structure on the birds in a tropical savanna woodland in north-eastern Australia', *Rangelands Journal* 28, pp. 139–152

Taylor, JA & Tulloch, D 1985, 'Rainfall in the wet-dry tropics: extreme events at Darwin and similarities between years during the period 1870–1983', *Australian Journal of Ecology* 10, pp. 281–295.

Thackway, R & Cresswell, ID 2005, *An interim bogeorgraphic regionalisation for Australia*, Australian Nature Conservation Agency, Canberra.

Torgersen, T, Luly, J, de Dekker, P, Jones, MR, Searle, DE, Chivas, AR & Ullman, WJ 1988. 'Late Quaternary environments of the Carpentaria Basin, Australia', *Palaeogeography, Palaeoclimatology, Palaeoecology* 67, pp. 245–261.

van der Kaars, WA 1991, 'Palynology of eastern Indonesian marine piston-cores: a Late Quaternary vegetational and climatic record for Australasia', *Palaeogeography, Palaeoclimatology, Palaeoecology* 85, pp. 239–302.

Veevers, JJ 2000, *Billion-year earth history of Australia and its neighbours in Gondwanaland*, GEMOC Press, Macquarie University, Sydney, NSW

Voris, HK 2000, 'Maps of Pleistocene sea levels in Southeast Asia: shorelines, river systems and time durations', *Journal of Biogeography* 27, pp. 1153–1167.

Walker, BH, Langridge, JL & McFarlane, F 1997, 'Resilience of an Australian savanna grassland to selective and non-selective perturbations', *Australian Journal of Ecology*, pp. 22:125–135.

Wheeler, JR, Rye, BL, Koch, BL & Wilson, AJG (eds) 1992, *Flora of the Kimberley region*, Western Australian Herbarium, Perth.

Whitehead, P 1999, 'Promoting conservation in landscapes subject to change: lessons from the Mary River', *Australian Biologist* 12, pp. 50–62.

Whitehead, P & Dawson, T 2000, 'Let them eat grass', *Nature Australia* Autumn 2000, pp. 46–55.

Whitehead, PJ, Wilson, BA & Bowman, DMJS 1990, Conservation of coastal wetlands of the Northern Territory of Australia: the Mary River floodplain', *Biological Conservation*, vol. 52, pp. 85–111.

Whitehead, PJ, Wilson, BA & Saalfeld, K 1992, 'Managing the magpie goose in the Northern Territory: approaches to conservation of mobile fauna in a patchy environment', *Conservation and development issues in northern Australia*, eds I Moffatt and A Webb, North Australia Research Unit: Darwin, pp. 90–104.

Williams, RJ, Woinarski, JCZ & Andersen, AN 2003, 'Fire experiments in northern Australia: lessons for ecology, management and biodiversity conservation', *International Journal of Wildland Fire* 12, pp. 391–402.

Williams, RJ, Hutley, LB, Cook, GD, Russell-Smith, J, Edwards, A & Chen, X 2004, 'Assessing the carbon sequestration potential of mesic savannas in the Northern Territory, Australia: approaches, uncertainties and potential impacts of fire', *Functional Plant Biology* 31, pp. 415–422.

Williams, RJ, Carter, J, Duff, GA, Woinarski, JCZ, Cook, GD & Farrer, SL 2005, 'Carbon accounting, land management, science and policy uncertainty in Australian savanna landscapes: introduction and overview', *Australian Journal of Botany* 53, pp. 583–588.

Woinarski, JCZ 2000, 'The conservation status of rodents in the monsoonal tropics of the Northern Territory', *Wildlife Research* 27, pp. 421–435.

Woinarski, JCZ, Brennan, K, Cowie, I, Fisher, A, Latz, PK & Russell-Smith, J 2000a, 'Vegetation of the Wessel and English Company islands, north-eastern Arnhem Land, Northern Territory, Australia', *Australian Journal of Botany* 48, pp. 115–141.

Woinarski, JCZ, Brock, C, Armstrong, M, Hempel, C, Cheal, D & Brennan, K 2000b. 'Bird distribution in riparian vegetation of an Australian tropical savanna: a broad-scale survey and analysis of distributional data base', *Journal of Biogeography* 27, pp. 843–868.

Woinarski, JCZ, Milne, DJ & Wanganeen, G 2001, 'Changes in mammal populations in relatively intact landscapes of Kakadu National Park, Northern Territory, Australia', *Austral Ecology* 26, pp. 360–370.

Woinarski, JCZ & Ash, AJ 2002, 'Responses of vertebrates to pastoralism, military land use and landscape position in an Australian tropical savanna', *Austral Ecology* 27, pp. 311–323.

Woinarski, JCZ, Andersen, AN, Churchill, TB & Ash, A 2002, 'Response of ant and terrestrial spider assemblages to pastoral and military land use and to landscape position, in a tropical savanna woodland in northern Australia', *Austral Ecology* 27, pp. 324–333.

Woinarski, JCZ, Brennan, K, Cowie, I, Kerrigan, R & Hempel, C 2003, *Biodiversity conservation on the Tiwi islands, Northern Territory. Part 1. Plants and environments*, Department of Infrastructure Planning and Environment: Darwin, pp. 144.

Woinarski, JCZ, Risler, J & Kean, L 2004a, 'The response of vegetation and vertebrate fauna to 23 years of fire exclusion in a tropical *Eucalyptus* open forest, Northern Territory, Australia', *Austral Ecology* 29, pp. 156–176.

Woinarski, JCZ, Armstrong, M, Price, O, McCartney, J, Griffiths, T & Fisher, A 2004b, 'The terrestrial vertebrate fauna of Litchfield National Park, Northern Territory: monitoring over a 6-year period and response to fire history', *Wildlife Research* 31, pp. 1–10.

Woinarski, JCZ, Williams, RJ, Price, O & Rankmore, B 2005, 'Landscapes without boundaries: wildlife and their environments in northern Australia', *Wildlife Research* 32, pp. 377–388.

Woinarski, JCZ, Hempel, C, Cowie, I, Brennan, K, Kerrigan, R, Leach, G & Russell-Smith, J 2006a, 'Distributional patterns of plant species endemic to the Northern Territory, Australia', *Australian Journal of Botany* 54, pp. 627–640.

Woinarski, JCZ, McCosker, JC, Gordon, G, Lawrie, B, James, C, Augusteyn, J, Slater, L & Danvers, T 2006b, 'Monitoring change in the vertebrate fauna of central Queensland, Australia, over a period of broad-scale vegetation clearance, 1975–2002', *Wildlife Research* 33, pp. 263–274.

Yibarbuk, D, Whitehead, PJ, Russell-Smith, J, Jackson, D, Godjuwa, C, Fisher, A, Cooke, P, Choquenot, D & Bowman, DMJS 2001, 'Fire ecology and Aboriginal land management in central Arnhem Land, northern Australia: a tradition of ecosystem management', *Journal of Biogeography* 28, pp. 325–343.

INDEX

Figures in **bold** refer to photographs

A

Aboriginal **4**, 8, 9, 16, 20, 21, 22, 24, 25, 27, 34, 35, 39, 40, 48, 64, 66, 67, 68, 71, 72, 75, 76, 81, 82, 83, 89, 91, 92, 93, 94, 95, 97, 98, 103, 106

Aboriginal Lands 25, 35, 67, 82, 103

abundance 1, 9, 14, 15, 16, 18, 19, 21, 34, 35, 39, 40, 42, 45, 46, 55, 66, 67, 68, 72, 74, 75, 78, 85

Acacia nilotica 78

Africa 7, 9, 13, 14, 16, 24, 47, 50, 61, 76, 77, 78, 81

agriculture vii, 7, 21, 25, 27, 29, 55, 76, 78, 101, 108, 113, 114

Allosyncarpia ternate 46

animals 1, 2, 7, 8, 9, 11, 14, 15, 16, 17, 18, 27, 31, 33, 34, 35, 37, 38, 39, 41, 42, 43, 45, 46, 47, 48, 55, 56, 58, 60, 64, 66, 67, 68, 72, 73, 76, 77, 78, 81, 82, 86, 89, 91, 92, 97, 98, 101, 109, 111, 114. *See also* wildlife; *See also* fauna; *See also* feral animals

Anindilyakwa archipelago 97

ants 16, 19, **19**, 39, 41, 42, 74, 76

aquaculture 58, 101

aquifer 14, 15, 31

Arafura plain 8, 9

Arafura Sea 3, 7, 42

arid 3, 9, 19, 24, 55, 59. *See also* semi-arid

Arnhem Land 8, 9, 11, 14, 15, 16, 21, 24, 25, 32, 37, 42, 45, 46, **46**, 47, 61, 64, 66, 71, 75, 76, 81, 91, 94, 95, **95**

Asia 9, 13, 19, 21, 23, 24, 41, 47, 77, 78, 83

Australian Bustard 40

B

Banded Fruit-dove 46

bandicoots 35, 37, 59, 64, 66

barramundi ii, 30, 41, 55

Bathurst Island 43

bee-eaters 41

biodiversity 4, 5, 15–18, 19, 20, 27, 31, 45, 49, 56, 58, 59–60, 63, 64, 65, 66, 69, 70, 71, 74, 77, 81, 82, 83, 85, 86, 88, 90, 91, 92, 94, 97, 98, 100, 103, 104, 105, 106, 107

biological diversity 13, 64

birds **vi**, 1, 16, 17, 18, **18**, 31, 37, 38, 39, 40, 41, 43, 45, 46, 47, **47**, **50**, 55, 58, 60, 64, 67, 68, 74, 75, 76, 77, **84**, 97, 110, 111

Black-faced Wood-swallow 68

Black-footed Tree-rat 66

Black-throated Finch 68

Black Cockatoo 17, 41. *See also* Red-tailed Black Cockatoo

Black Kite 18, 41

Black Wallaroo 46

boabs 7, 15, 16

Brush-tailed Phascogale 64, **69**

Brush-tailed Possum 35, 66, **99**

Brush-tailed Rabbit-rat 64, 97

buffalo 46, 48, **62**, 67, 71, 73, 74, 75, 97, 111, 114

Buffel Grass 79

Burrowing Bettong 64

Bush Stone-curlew 40

bush tucker 38, 40, 75, 76, 98

C

Cabomba caroliniana 78

Calotropis procera 78

cane toads 76, **76**, 97

Cape York **inside front cover, ii, vi, viii**, 2, 3, **4**, **5**, 7, 8, 9, 10, 11, 12, 13, 14, 15, **15**, 16, 18, 20, **20**, 21, 22, 23, **23**, 25–27, **30**, 33, **36**, **38**, **40**, 42, 43, **44**, 45, **48**, 50, **51**, **57**, 61, 68, 69, 76, 93, **106**, **112**

carbon 82, 91, 92

  carbon economy 92

  carbon trading 4, 92, 106

catchments 3, 33, 55–58, 59, 70, 71, 103, 116

Cathormium umbellatum 33

cats 43, 72, 75, 76

cattle **ii**, 14, 21, **21**, 22, 24, 26, 27, 34, 39, 48, 67, 68, 70, 71, 72, 73, **73**, 74, 78, 81, 92, 97, 100, **100**, **102**, 103, 110, 113, 114

channel country 55

Chestnut-quilled Rock-pigeon 46

climate 2, 3, 8, 9, 11, 12, 14, 16, 21, 23, 27, 29, 33, 34, 39, 42, 46, 47, 49, 69, 82–83, 88, 91, 94, 107

Cobourg Peninsula 64, 73

conservation 3, 4, 5, 22, 25, 35, 39, 42, 45, 47, 50, 56, 61, 66, 67, 68, 73, 75, 81, 85, 86, 87, 88, 89, 90, 92, 93, 95, 97, 98, 99, 101, 102, 103, 104, 105, 106, 107, 108, 109

  conservation planning 85–88, 89

  conservation reserves. *See* reserves

  conservation values 27, 29, 75, 82, 86

Cooktown 3, 14, 22

Cooktown Ironwood 12, 24

Coral Sea 3

Crested Pigeons 73

crocodiles 33, 41, 43, 82, 98

  saltwater crocodiles 14, 15, **38**, 58

cropping 21, 22, 25, 48, 52, 104, 113, 115, 116. *See also* agriculture

crustaceans 46

cultural 4, 9, 20, 27, 33, 34, 46, 58, 71, 89, 92, 93, **93**, 94, 96, 97, 98, 103, 109

customs 4, 20, 94, 98

cyclones 9, 30, 34, 42, 82, 107

cypress-pines 12, 37, 72

THE AUTHORS

Dr John Woinarski is a Principal Scientist with the Northern Territory Department of Natural Resources Environment and the Arts, and an Adjunct Professor at Charles Darwin University. Dr Woinarski is recognised internationally as an expert in conservation ecology, especially the ecology of Northern Australia. Dr Woinarski has spent the last 20 years conducting scientific research on conservation and management issues in Northern Australia, and has published more than 150 scientific research papers on a wide range of conservation issues. In 2001, Dr Woinarski was awarded Australia's premier science award, the Eureka Prize for his ecological research.

Professor Brendan Mackey is a professor of environmental science in the Fenner School of Research & Environment, College of Science, at The Australian National University, and is Director of the ANU WildCountry Research & Policy Hub. His doctoral thesis was on tropical forest ecology in far North Queensland. Professor Mackey is currently lead chief investigator for an Australian Research Council grant researching connectivity conservation issues in Northern Australia. Professor Mackey was the lead author of a major report for the Queensland government on the natural heritage significance of Cape York Peninsula. He has worked as a research scientist with CSIRO and the Canadian Forest Service, and is involved in various international programs through the IUCN (World Conservation Union) and the Earth Charter initiative. Professor Mackey has written over 100 scientific publications on environmental matters.

Emeritus Professor Henry Nix AO is co-chair of the Australian Wild Country Science Council, with Emeritus Professor Michael Soulé. Professor Nix is a Visiting Fellow and former Director, from 1986–1999, of the Centre for Resource and Environmental Studies (CRES), now the Fenner School of Environment and Society, at the Australian National University. Prior to this he was Senior Principal Research Scientist in the CSIRO Division of Land and Water. He is a pioneer of computer-based methods of land resource inventory and evaluation for a wide range of potential uses that include field and horticultural crops, forestry, hydrology and biodiversity conservation. His continent-wide evaluation of the climate, terrain and soil constraints on arable land revealed the extremely limited potential of Northern Australia for agricultural development. He has travelled extensively across all regions of the North, spending four to six weeks each year since 1985 in field surveys of freshwater fish and riparian birds. He is a Life Member, former Councillor and President of Birds Australia 2001–2005.

Dr Barry Traill is currently Director of the Wild Australia Program of the Pew Charitable Trusts. Prior to this he worked for 30 years as an ecologist. Dr Traill has worked as a conservation advocate with The Wilderness Society and other government, non-government and business organisations, specialising in the conservation of tropical and temperate woodlands. He has initiated and led a wide range of conservation programs in woodland and savanna areas in northern and southern Australia. He has particular expertise in the ecology of woodland birds and mammals. He is President of the Invasive Species Council of Australia.

www.ingramcontent.com/pod-product-compliance
Lightning Source LLC
Chambersburg PA
CBHW051301270326
41926CB00030B/4690